TK 5103.5 .M56 c.5

RANDOM SIGNAL ANALYSIS

DWIGHT F. MIX
University of Arkansas

RANDOM SIGNAL ANALYSIS

ADDISON-WESLEY PUBLISHING COMPANY
Reading, Massachusetts · Menlo Park, California · London · Don Mills, Ontario

This book is in the
ADDISON-WESLEY SERIES IN ELECTRICAL ENGINEERING

Consulting Editors
DAVID K. CHENG
LEONARD A. GOULD
FRED K. MANASSE

Copyright © 1969 by Addison-Wesley Publishing Company, Inc. All rights reserved. No part of this publication may be reproduced, stored in a retrieval system, or transmitted, in any form or by any means, electronic, mechanical, photocopying, recording, or otherwise, without the prior written permission of the publisher. Printed in the United States of America. Published simultaneously in Canada. Library of Congress Catalog Card No. 69-11630.

To Clyde Jackson Stoker

PREFACE

Electric circuit analysis courses are continually changing. Ten years ago many schools required only steady-state analysis in the undergraduate curriculum. Now steady-state analysis is taught as a special case of the more general theory.

Today many schools require only deterministic analysis, though some are beginning to introduce random signal theory to undergraduates. Of course, deterministic signal theory may never be taught as just a special case of the more general theory, but random signal theory will surely enter the undergraduate curriculum of most universities in the near future.

This text was written for use in a required circuits course taught at the senior level at the University of Arkansas. The necessary background for the course is circuit analysis through the Laplace transform.

The text is divided into three parts, with the last two containing the material on random signals. Part 1 is the background material in Fourier methods necessary to understand parts 2 and 3. A typical course would contain a quick review of part 1 and then proceed to a detailed study of parts 2 and 3. There is adequate material here for a three-hour semester course.

I have been teaching such a course for the past three semesters, using rough draft copies of this text. Since the senior students here at the University of Arkansas have an adequate background in Fourier analysis but no knowledge of probability theory, we begin in Chapter 5 and continue to the end of the text. This is still a bit more material than one can conveniently cover in one semester. The text should be just about the right length for courses in those schools in which the students have an adequate background in both Fourier analysis and probability theory.

I would like to acknowledge the influence that my association with Purdue University has had on this work, particularly by Dr. John C. Hancock and Dr. Ronald E. Totty (now with Radiation Inc., Melbourne, Florida). The Electrical Engineering Department at the University of Arkansas, headed by Dr. Denys O. Akhurst, has a most generous policy toward aspiring authors. For this I am grateful. Also special thanks are due to Mrs. Patricia Edge for her careful typing of the manuscript.

Fayetteville, Arkansas   D.F.M.
August 1968

CONTENTS

PART 1 **Functions and Operators: The Time Domain and the Frequency Domain**

Chapter 1 **Introduction**

1.1	Background	3
1.2	Classification of signals	5

Chapter 2 **Fourier Series**

2.1	The Fourier series as an operator	10
2.2	The Fourier series as an orthogonal expansion	13
2.3	Expansion of an arbitrary signal	16
2.4	Fourier series expansion of periodic time functions . . .	17
2.5	Properties of Fourier series	21
2.6	Use of the differentiation property to evaluate coefficients . .	23
2.7	Parseval's theorem and power spectral densities	25
2.8	Examples	28

Chapter 3 **The Fourier Transform**

3.1	The Fourier transform as an operator	39
3.2	Relationship between the Fourier transform and other operators	41
3.3	Properties of the Fourier transform	44
3.4	Use of the Fourier transform properties	44
3.5	Bounds on the spectrum	47
3.6	Rayleigh's theorem and energy spectral density	49
3.7	Examples	51

Chapter 4 **Systems**

4.1	Classification of systems	61
4.2	Types of response	64
4.3	The system function $H(s)$ in the frequency domain	66
4.4	The system function $h(t)$ in the time domain	69
4.5	Relationship between differential equations and the impulse response	74

PART 2 Probability and Random Variables

Chapter 5 Probability

    5.1 Some elementary set theory 85
    5.2 Sets and events 90
    5.3 Discrete and continuous sample spaces 93
    5.4 Partitioning 94
    5.5 Conditional probability 95
    5.6 Mutually exclusive and independent events 98
    5.7 Examples 100

Chapter 6 Random Variables

    6.1 Definition of random variable 107
    6.2 Distribution of random variables 108
    6.3 Relationship to electrical engineering 112
    6.4 Statistical averages 117
    6.5 Some important distributions 120
    6.6 Characteristic functions 127

Chapter 7 Random Variables (*Continued*)

    7.1 Function of a random variable 134
    7.2 Determination of the cdf and pdf of $Y = g(X)$ 135
    7.3 Examples for finding the distribution of $Y = g(X)$ . . . 139
    7.4 Jointly distributed random variables 144
    7.5 Conditional probability distribution 149

Chapter 8 Limit Theorems and the Normal Distribution

    8.1 Limit theorems 157
    8.2 The gaussian random variable 164
          Appendix: Matrices 173

PART 3   Stochastic Processes and Random Signals

Chapter 9   Stochastic Processes

    9.1   Definition of a stochastic process 183
    9.2   Moments of stochastic processes 185
    9.3   Specifying stochastic processes 189
    9.4   The gaussian process 190
    9.5   Stationary and ergodic processes 191

Chapter 10   The Power Spectrum and its Relation to the Autocorrelation Function

    10.1   Power spectral density and autocorrelation 199
    10.2   Calculating moments of stochastic processes specified by the transformation of variables 202
    10.3   Calculating moments of stochastic processes specified by operators (systems) 214

Chapter 11   Mean-Square Estimation

    11.1   Introduction 223
    11.2   Mean-square estimation 224
    11.3   Vectors and linear mean-square estimation 229
    11.4   Continuous data 234

Chapter 12   Optimum Systems

    12.1   Introduction 239
    12.2   The optimum type I filter 240
    12.3   The optimum type II filter 244
    12.4   Prediction in the absence of noise 249

           Table of Normal Curve Areas 257

           Answers to Selected Problems 261

           Index 269

PART 1

Functions and Operators:
The Time Domain and the Frequency Domain

1
INTRODUCTION

1.1 BACKGROUND

In the problems considered in basic circuit theory courses the input signal is some function of time $v_1(t)$, and the output, say $v_2(t)$, is related to the input by a system function $h(t)$. In the frequency domain, the output of a linear system is often found by transform techniques, and in the time domain, by the solution of the system differential equation. The output can also be found in the time domain by convolution methods. A linear system is pictured in Fig. 1.1. A salient feature of this system is that we can write an equation for the input $v_1(t)$.

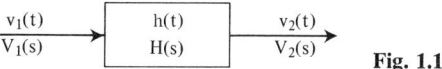

Fig. 1.1

Suppose the input $v_1(t)$ is a random signal such as a voice waveform. The signals received by a table radio, a TV signal, and police communication are examples of random waveforms for which it is difficult to write equations. Furthermore, even if the past history of such a waveform could be adequately described by an equation, the future would be unpredictable, and therefore it would be impossible to describe the entire waveform by an equation.

Since an equation that describes the input cannot be written, it is impossible to find the output as a function of time.

What techniques are available for the analysis of systems with random inputs? In the early days of radio, engineers circumvented this problem by assuming that the input was deterministic (usually a sinusoid), and then the system was designed for this input signal. Although this technique was quite successful, it amounted simply to sidestepping the problem. Furthermore, assuming a deterministic signal is not useful in all problems. Therefore a general method for handling random signals is needed.

Naval warfare is an example of a problem that cannot be solved by assuming a deterministic input. Big naval guns can propel a projectile over the horizon (about 26 miles), and the time between when the gun is fired and when the projectile reaches its target can exceed one minute. During this minute the target (an enemy ship) can move completely out of the target area. Therefore it becomes necessary to predict the future position of the enemy ship.

In this case the input to our system is the record of the enemy ship's positions in the past. The system itself can be an electronic device whose output is an estimate of the enemy ship's position one minute from the time of the last position reported. This electronic device can then be used to direct and fire the gun.

This problem was one of the earliest applications of Wiener filtering and prediction theory, first published in 1939 by Norbert Wiener at M.I.T. At about the same time the Russian radio engineer Kotel'nikov published his work in which he arrived independently at the same conclusions, although his work was not generally known in this country until after the war. The work of these two men forms the basis for the solution of one aspect of the analysis problem, and we will consider their work in Chapter 12 of this text.

Let us again consider the analysis problem in which the input is deterministic, and let us see what can be solved in this case. Since the output can be found explicitly as a function of time, we are content to simply find this time function, without asking why. However, this is not possible when we have a random input. Perhaps we can solve our problem directly without finding the output time function if we consider out ultimate aim. Generally, our ultimate aim is to design an optimum system, and the step of expressing the system output as a time function is only an intermediate step. Perhaps there is a better (and more general) way to design our system than the old procedure in which the output must be expressed explicitly. Of necessity, this is the approach taken in all problems in which the input is random.

As an example, consider the commercial radio system in use today, and suppose we wish to design a good receiver. By "a good receiver" we mean one that will reject as much extraneous noise as possible while amplifying the signal. If we can determine the frequency spectrum of the received signal and of the noise, then we can design our receiver as a filter to reject noise and amplify signal. The Wiener filtering and prediction theory expresses this philosophy in mathematical form. (It is interesting to note that the early design of commercial receivers was very nearly optimum, as we will see in Chapter 12.)

Another problem is the design of optimum digital communications systems, such as commercial teletype. A typical system transmits a sequence of binary on-off pulses. If no noise were added to the received signal, the sequence of binary pulses could be decoded without error. But noise is always present in electronic devices, and since signal energy decreases as the distance between transmitter and receiver increases, errors become unavoidable at large distances. The proper design of such communications systems is a major problem to the communications engineer, and he must know the basic theory of random signals before tackling this problem.

Current efforts in oil exploration also draw upon the theory of random signal analysis. The diamond-studded drill bits used to bore through hard rock cost something like $10,000 each, and often several of these are used to drill through a single layer of rock. Therefore it is understandable that efforts should be made to ensure the presence of oil before the digging operation begins. To conduct preliminary exploration, one detonates explosive charges on the earth's surface. Sound waves are reflected from the various layers in the earth's crust, and this acoustic information can be analyzed to determine whether oil-bearing rock is present. However, the solution is not quite this simple, since reflected waves are received laterally as well as from within the earth's crust. Techniques for distinguishing the desired signal from the unwanted (lateral) reflections have been devised, and these techniques involve correlation, which we will discuss in Chapter 10.

A closely related problem is the detection of submarines. Unfortunately, the use of sonar announces the presence of the detecting ship as surely as it detects the enemy submarine. Techniques that will enable a ship to detect enemy submarines without emitting radiation are being sought. As a submarine moves through water it emits sound waves that are similar to those made by large fish. It then becomes difficult to separate the submarine's noise from the noise of the fish and other sea organisms. This is an interesting and complicated problem.

The solution of problems such as we have discussed draws on such diverse fields as Fourier and Laplace transform theory, probability theory, statistics, and stochastic processes. In Part 1 of this text we consider Fourier theory, in Part 2 we discuss probability theory, and in Part 3 we combine the concepts of Parts 1 and 2 to solve some of the problems discussed above. Hopefully, a broad enough view of random signal analysis will be presented in the succeeding chapters so that the reader can gain a basic understanding of the theory.

1.2 CLASSIFICATION OF SIGNALS

Signals may be classified into one of two broad categories:
 1. power signals,
 2. energy signals.

  Power signals may then be classified as
1a. periodic,
1b. almost periodic,
1c. random,
and energy signals may be classified as
2a. deterministic,
2b. random.

It is easy to find fault with this classification system (for example, a periodic signal may have random phase), but it is good enough for our purposes.

Examples of power signals are the voice or music waveforms in commercial radio, teletype signals, and any periodic voltage, such as the 60-cycle commercial power voltage. This 60-cycle voltage is called a power signal because it fits our definition of "power" (see Definition 1), not because it supplies power to our homes. These signals can also be considered energy signals, rather than power signals, depending on how they are defined.

An energy signal is a pulselike signal that usually exists for only a finite time. Therefore, if any of the above signals are considered to exist only for a finite time, they become energy signals rather than power signals. Thus the primary difference between power and energy signals is that a power signal has been present from the infinite past and will be present into the infinite future. Although an energy signal can be present for an infinite time, the major portion (most of the area) of an energy signal is concentrated in a finite time interval.

All signals can be classified into one of the two categories. Power signals have nonzero but finite power. Energy signals have nonzero but finite energy. We now formalize these concepts.

Definition 1. *Power.* The power in a signal $f(t)$ is given by

$$P = \lim_{a \to \infty} \frac{1}{2a} \int_{-a}^{+a} |f(t)|^2 \, dt. \tag{1.1}$$

Notes. a) This definition follows from the classical or circuit theory definition of *average* power. If we identify our function as current, $f(t) = i(t)$, then the instantaneous power is given by

$$p(t) = i^2(t)R,$$

and the average power is defined as

$$P_{av} = \lim_{a \to \infty} \frac{1}{2a} \int_{-a}^{+a} i^2(t)R \, dt.$$

If $R = 1 \, \Omega$ (ohm), then Eq. (1.1) corresponds to the usual (circuit theory) definition of average power.

b) If $f(t)$ is a real function of time, then $|f(t)|^2 = f^2(t)$. If $f(t)$ is complex, then $|f(t)|^2 = f(t) \cdot f^*(t)$, where $f^*(t)$ denotes the complex conjugate of $f(t)$. Of course, if $f(t)$ is periodic, then Eq. (1.1) reduces to

$$P = \frac{1}{T} \int_0^T |f(t)|^2 \, dt.$$

1.2 CLASSIFICATION OF SIGNALS

Definition 2. *Energy.* The energy in a signal $f(t)$ is given by

$$E = \int_{-\infty}^{+\infty} |f(t)|^2 \, dt. \tag{1.2}$$

Notes. a) This is for *total* energy, just as Definition 1 was for *average* power (on a one-ohm basis).

b) The instantaneous power, $p(t)$, and instantaneous energy, $w(t)$, are related by

$$p(t) = \frac{d}{dt} w(t)$$

or, equivalently, by

$$w(t) = \int_{-\infty}^{t} p(\lambda) \, d\lambda,$$

where $p(t)$ and $w(t)$ are computed for the *same* signal $v(t)$.

There is no such simple relationship between average power P and total energy E, where P and E are computed for the same signal $v(t)$. In fact, if E is finite, then P is zero. If P is nonzero, then E is infinite.

c) Other names for energy signals are *pulse signals* and *aperiodic signals*. Some examples of aperiodic signals are given in Fig. 1.2.

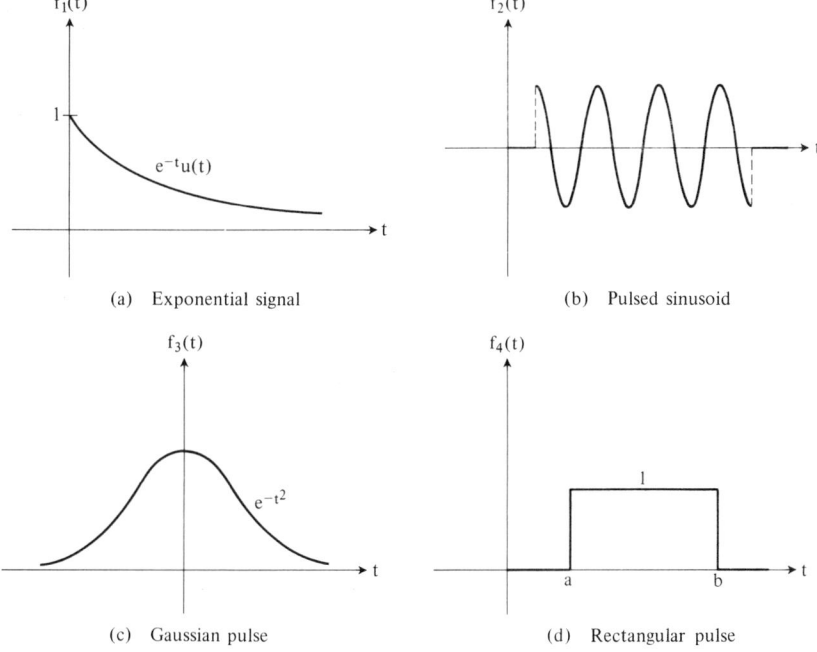

Fig. 1.2 Some energy signals.

Definition 3. *Periodic functions.* A function is periodic with period T if it repeats itself, that is, if

$$f(t) = f(t + T) \qquad \text{for all } t. \tag{1.3}$$

The fundamental period is the smallest T for which Eq. (1.3) holds.

A function of special importance to us is $f(t) = e^{jt}$. It is periodic with period $T = 2\pi$. The function $f(t) = e^{jt}$ may be represented by a rotating vector (phasor) that has magnitude one and angular position $\theta = t$ (Fig. 1.3). As time increases the phasor rotates counterclockwise through 2π radians. The position of the vector is the same at time t and at time $t + 2\pi$. This is the justification for the periodic representation of e^{jt}.

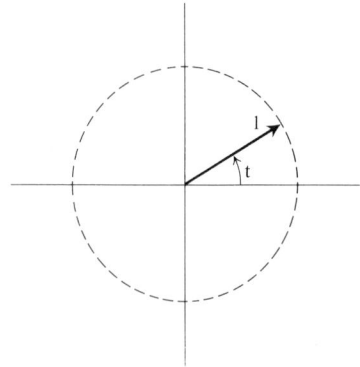

Fig. 1.3 The phasor e^{jt}.

A function is almost periodic if it can be approximated to an arbitrary degree of closeness by the sum of a finite number of periodic functions. Thus we have the following definition for almost periodic functions.

Definition 4. *Almost periodic functions.* A function $f(t)$ is almost periodic if

$$f(t) = \lim_{k \to \infty} \sum_{i=1}^{k} \phi_i(t),$$

where each $\phi_i(t)$, $i = 1, 2, \ldots$, is either periodic or zero.

Note. Every periodic function is almost periodic, but the converse does not hold. Consider

$$f(t) = \sin t + \sin \sqrt{2} t. \tag{1.4}$$

This function is almost periodic since each term on the right-hand side of Eq. (1.4) is periodic. It is not periodic since there is no period T in which $f(t)$ repeats itself.

Random signals can be classified as either power signals or energy signals. A random signal is one that cannot in general be described by a simple mathematical equation. The waveform that drives a radio speaker is an example of a random power signal. A pulse of unpredictable height is an example of a random energy signal.

A random signal is best described as a sample function from a random process. These concepts will be described in Chapter 9, so we will not attempt a definition of random signals here.

The above classification system will prove useful to us in the remainder of the text. Periodic power signals can be represented by Fourier series, and energy signals can be represented by Fourier integrals. Some special procedures must be developed before we can represent a random power signal in the frequency domain. The primary tool used here will be the Fourier series, and these procedures will be developed in Chapter 10. Throughout the text we can keep things in their proper place by noting whether we are dealing with a power signal or an energy signal.

PROBLEMS

1. Identify each of the following signals as either a power signal or an energy signal.
   a) $v(t) = \sin \omega t$, $-\infty < t < +\infty$
   b) $v(t) = \begin{cases} 10e^{-t}, & t > 0, \\ 0, & t < 0 \end{cases}$
   c) The signal shown in Fig. 1.4
   d) The waveform of (c), periodic with period $T = 2$
2. Find the power and energy in each of the signals in Problem 1.

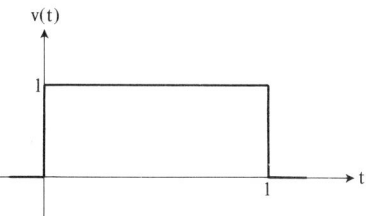

Fig. 1.4

FURTHER READING

1. S. J. MASON and H. J. ZIMMERMAN, *Electronic Circuits, Signals, and Systems*, Wiley, New York, 1960.
2. B. P. LATHI, *Signals, Systems, and Communication*, Wiley, New York, 1965.

The material in Section 1.2 on power signals and energy signals is discussed in Chapter 6 of Mason and Zimmerman and in Sections 3.9 and 4.9 of Lathi.

2
THE FOURIER SERIES

2.1 THE FOURIER SERIES AS AN OPERATOR

The Fourier series is an operator. Other operators that we shall deal with are the Fourier integral and the Laplace transform. An operator is a special type of function.

The reader will recall that a function is a relationship between two sets (called the domain and the range), and this relationship must satisfy two conditions. First, every element in the domain must correspond to some element in the range. Second, no two elements in the range can correspond to the same element in the domain. The major point here is that the elements in these two sets (domain and range) are unspecified. Usually the elements of the sets are numbers, but they can be animals, books, or even functions.

Children taking the "new math" in grade school are introduced to the concept of function in the following way. A "function machine" is described as a computer that produces an output by performing a certain operation on the input (Fig. 2.1). Thus Function machine A adds 2 to the input. A number x is selected from the domain and fed into the machine. The machine associates a number $y = x + 2$ with this number x. Similarly, Function machine B associates the number $y = 3x$ with each x in the domain.

An operator is a function whose domain and range are both sets of functions. In terms of the function machines pictured in Fig. 2.1, an operator is simply a function machine in which the input x is a function, say $x(t)$, and

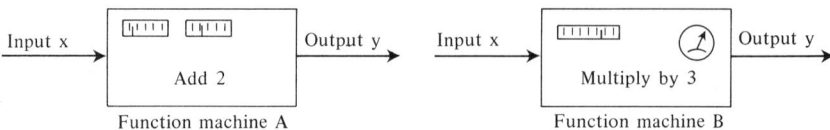

Fig. 2.1 Function machines.

the output y is also a function, say $y(t)$. (It is not necessary for the input and output to be functions of the same variable.)

Note that our mathematical model for a system is an operator. See Fig. 1.1.

The Fourier series operator is the relationship between $v(t)$ and $V(f_k)$ in the following equation:

$$V_k = V(f_k) = \frac{1}{T} \int_{t'}^{t'+T} v(t)e^{-j2\pi f_k t} \, dt. \tag{2.1}$$

It is shown in Fig. 2.2. The function machine "forward operator" performs the indicated operations on a particular function v to provide the output V.

Now v and V are themselves functions. The input v is a function of the continuous parameter t (by this we mean that values of t are in the domain of v), and V can be thought of as a function of either the index k or the discrete parameter f_k, in which case we write $V(f_k)$. Since we are concerned with signals as functions of time, we will identify t as time. The parameter f_k represents a discrete frequency.

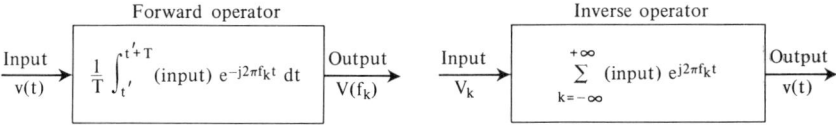

Fig. 2.2 The function machine "forward operator."

Fig. 2.3 The function machine "inverse operator."

Note. The parameter t can represent any physical quantity or attribute. For example, one application has identified t as space and f_k or k as wave number. The French mathematician Fourier first used the series in his study of heat—an application which is physically quite different from ours.

In the inverse operation the domain is a set of functions $\{V\}$ and the range is a set of functions $\{v\}$. This inverse operation is

$$v(t) = \sum_{k=-\infty}^{+\infty} V_k e^{j2\pi f_k t}. \tag{2.2}$$

Again, this is an operator; it is illustrated in Fig. 2.3. The input to our function machine "inverse operator" is a function $V(f_k) = V_k$, and the output is a function $v(t)$.

Specifying the Domain

A function is specified when the formula (mapping) and the domain are both given. Since an operator is just a special type of function, if we wish to use the operators given in Eqs. (2.1) and (2.2), we must specify the domain.

The domain of the forward operator is given by

$$\int_{t'}^{t'+T} |v(t)|^2 \, dt < \infty. \tag{2.3}$$

That is, any function $v(t)$ that satisfies Eq. (2.3) is in the domain of the forward operator, and may therefore be used in Eq. (2.1).

The domain of the inverse operator is given by

$$\sum_{k=-\infty}^{+\infty} |V_k|^2 < \infty, \tag{2.4}$$

and any function V_k that satisfies this inequality may be used in Eq. (2.2).

Notes. a) Any function $v(t)$ that satisfies Eq. (2.3) has a Fourier series whose coefficients V_k satisfy Eq. (2.4). Conversely, if a function V_k satisfies Eq. (2.4), then the corresponding $v(t)$ satisfies Eq. (2.3).

b) Convergence in the mean is guaranteed by Eq. (2.3). Any function $v(t)$ that satisfies Eq. (2.3) has a Fourier series that converges in the sense that

$$\int_{t'}^{t'+T} \left| \sum_{k=-n}^{+n} V_k e^{j2\pi f_k t} - v(t) \right|^2 dt \to 0 \quad \text{as} \quad n \to \infty.$$

This does *not* mean that the sum is equal to $v(t)$ at every t. For instance, if $v(t)$ has a jump discontinuity at some $t = t'$, the sum converges to a value between the right- and left-hand limits.

c) Any signal that can be generated in the laboratory satisfies the condition of Eq. (2.3).

Examples

1. Suppose that $v(t)$ is a sinusoid:

$$v(t) = E \sin \omega_a t. \tag{2.5}$$

Then $v(t)$ has a Fourier series, for Eq. (2.3) gives

$$\int_{t'}^{t'+T} E^2 \sin^2 \omega_a t \, dt = \frac{E^2 T}{2} < \infty.$$

2. The Fourier series for the sine wave of Eq. (2.5) is

$$V_{-1} = -\frac{E}{2j},$$

$$V_1 = \frac{E}{2j},$$

$$V_k = 0 \quad \text{for all other } k.$$

Testing this function in Eq. (2.4), we obtain

$$\sum_{k=-\infty}^{+\infty} |V_k|^2 = \frac{E^2}{2} < \infty.$$

Therefore the function V is in the domain of the inverse operator, Eq. (2.2), and this inverse operator will yield the sinusoidal function $v(t)$ of Eq. (2.5).

3. As stated before, any signal that can be generated in the laboratory is in the domain of the forward operator. Therefore, in order to obtain a function that does not satisfy Eq. (2.3), an absurd function must be "thought up." The function g described below is such a function:

$$g(t) = \begin{cases} 1, & t \text{ rational,} \\ 0, & t \text{ irrational.} \end{cases}$$

The reason that g does not satisfy Eq. (2.3) is related to the definition of an integral, which we need not go into here.

Any "reasonable" function satisfies Eq. (2.3), and therefore has a Fourier series. This example illustrates one type of "unreasonable" function that does not have a Fourier series.

2.2 THE FOURIER SERIES AS AN ORTHOGONAL EXPANSION

We have classified the Fourier series as an operator. It is also an orthogonal expansion. The orthogonality of the Fourier series allows us to derive many of its useful properties; in particular, it allows us to evaluate the coefficients with ease.

The meaning of the term "orthogonal" is well known when one is speaking of vectors. The meaning of the term "orthogonal functions" may not be clear, unless one recognizes that a function is a vector. This brings up the question "What is a vector?"

Although some vectors are "directed magnitudes," not all vectors are. Furthermore, not all directed magnitudes are vectors. For example, an elephant moving north is a directed magnitude, but not a vector.

Vectors are objects that have certain defining properties. Let us consider the properties of ordinary geometric vectors. Some of these properties are listed below; here X, Y, and Z represent vectors, and a, b, and c represent scalars.

1. The sum $X + Y + Z$ may be computed from the left or from the right; that is, $(X + Y) + Z = X + (Y + Z)$.
2. There is a zero vector, 0, with the property that $0 + X = X$ for all X.
3. The numerical multiple $(-1)X$ acts like the negative of X, since $(-1)X + X = 0$.

4. The sum $X + Y$ may be computed in any order; that is, $X + Y = Y + X$.
5. Vectors multiplied by the product of two scalars have the property that $a(bX) = (ab)X$.
6. Vectors multiplied by the sum of two scalars have the property that $(a + b)X = aX + bX$.
7. The sum of two vectors multiplied by a scalar has the property that $a(X + Y) = aX + aY$.

Any set of objects $X, Y, Z, \ldots$, that satisfy these seven properties is called a *vector space*. The objects themselves are called *vectors*. If, in addition to the above seven properties, a dot product (inner product) is defined for the vectors, the set is called an *inner product space*. Ordinary geometric vectors form an inner product space.

Now consider ordinary signals that we can generate in the laboratory as functions of time. These time functions $v_1(t), v_2(t), v_3(t), \ldots$, satisfy the seven conditions listed above. Hence they are vectors. If we define the inner product $\langle v_1, v_2 \rangle$ as

$$\langle v_1, v_2 \rangle = \int_{t_1}^{t_2} v_1(t) v_2(t)\, dt, \tag{2.6}$$

then these signals form an inner product space. We say that two signals are orthogonal (over the interval $t_1 < t < t_2$) if

$$\langle v_1, v_2 \rangle = 0. \tag{2.7}$$

The interval $t_1 < t < t_2$ is important. It is obvious, but worth stating, that two functions $v_1(t)$ and $v_2(t)$ may be orthogonal over one interval (t_1, t_2) but not over some other interval (t_3, t_4).

For real signals (such as those generated in the laboratory) Eq. (2.6) is a good definition of inner product. We will be concerned with the complex signal $e^{j\omega t}$, and therefore must extend the definition to complex signals.

Definition 1. *Inner product.* The inner product $\langle v_1, v_2 \rangle$ of two functions $v_1(t)$ and $v_2(t)$ is given by

$$\langle v_1, v_2 \rangle = \int_{t_1}^{t_2} v_1(t) v_2^*(t)\, dt, \tag{2.8}$$

where $v^*(t)$ represents the complex conjugate of $v(t)$.

Notes. a) If $v_1(t)$ and $v_2(t)$ are real, Eq. (2.8) reduces to Eq. (2.6).

b) The inner product is a functional.* Two functions $v_1(t)$ and $v_2(t)$ are fed into the function machine "inner product," and out pops a number $\langle v_1, v_2 \rangle$.

---

* A functional is a function with a domain of functions and a range of numbers.

2.2 THE FOURIER SERIES AS AN ORTHOGONAL EXPANSION

Use of the Orthogonality Property

The functions used in the Fourier series expansion are orthogonal to each other. Consider Eq. (2.2), rewritten here for convenience:

$$v(t) = \sum_{k=-\infty}^{+\infty} V_k e^{j2\pi f_k t}. \tag{2.2}$$

With $f_k = kf_1$ and $\omega = 2\pi f$, we rewrite Eq. (2.2) as

$$v(t) = \cdots + V_{-1}e^{-j\omega_1 t} + V_0 + V_1 e^{j\omega_1 t} + V_2 e^{j2\omega_1 t} + \cdots \tag{2.9}$$

The functions

$$g_k(t) = e^{jk\omega_1 t}, \quad k = \ldots, -2, -1, 0, 1, 2, \ldots$$

are orthogonal over the interval $t' < t < t' + T$, where $T = 1/f_1$. This is easily seen by

$$\langle g_k, g_l \rangle = \int_{t'}^{t'+T} e^{jk\omega_1 t} e^{-jl\omega_1 t} \, dt = \int_{t'}^{t'+T} e^{j(k-l)\omega_1 t} \, dt = \begin{cases} 0, & k \neq l, \\ T, & k = l. \end{cases}$$

This orthogonality of the functions $e^{j\omega_k t}$ is useful in evaluating the coefficients V_k in Eq. (2.2). As an example we will evaluate V_2.

Multiplying both sides of Eq. (2.9) by $e^{-j2\omega_1 t}$, we have

$$v(t)e^{-j2\omega_1 t} = \cdots + V_{-1}e^{-j3\omega_1 t} + V_0 e^{-j2\omega_1 t} + V_1 e^{-j\omega_1 t} + V_2 + V_3 e^{j\omega_1 t} + \cdots$$

Integrating both sides with respect to time over any interval of length T, we get

$$\int_{t'}^{t'+T} v(t)e^{-j2\omega_1 t} \, dt = \cdots + \int_{t'}^{t'+T} V_{-1}e^{-j3\omega_1 t} \, dt + \int_{t'}^{t'+T} V_0 e^{-j2\omega_1 t} \, dt$$

$$+ \int_{t'}^{t'+T} V_1 e^{-j\omega_1 t} \, dt + \int_{t'}^{t'+T} V_2 \, dt + \int_{t'}^{t'+T} V_3 e^{j\omega_1 t} \, dt + \cdots$$

$$= \cdots + 0 + 0 + 0 + V_2 T + 0 + \cdots$$

Solving for V_2 gives

$$V_2 = \frac{1}{T} \int_{t'}^{t'+T} v(t)e^{-j2\omega_1 t} \, dt,$$

or, for general k,

$$V_k = \frac{1}{T} \int_{t'}^{t'+T} v(t)e^{-jk\omega_1 t} \, dt,$$

which is Eq. (2.1).

Note. We have shown that the orthogonality of the vectors $e^{j\omega_k t}$ may be used to evaluate the coefficients V_k in the Fourier series. Furthermore, if a function $v(t)$ is written as a series, Eq. (2.2), the use of this orthogonality property leads naturally to Eq. (2.1), the forward operation on $v(t)$.

2.3 EXPANSION OF AN ARBITRARY SIGNAL $v(t)$

Let us choose some interval of length T and expand an arbitrary function $v(t)$ in terms of the complex exponentials $e^{j\omega_k t}$. Then, so long as Eq. (2.3) is satisfied, we can find the coefficients V_k by inserting the equation for $v(t)$ between t' and $t' + T$ into the operator, Eq. (2.1). See Fig. 2.4.

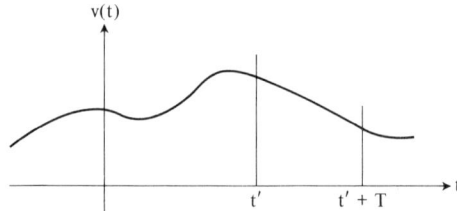

Fig. 2.4 An arbitrary function $v(t)$ to be expanded over the interval $(t', t' + T)$.

Having found the set of Fourier coefficients $\{V_k\}$ we may use Eq. (2.2) to obtain the time function $v(t)$ again. The function $v(t)$ will be given by

$$v(t) = \sum_{k=-\infty}^{+\infty} V_k e^{j2\pi f_k t} \qquad (2.2)$$

in the interval $(t', t' + T)$. Unfortunately, the set of numbers $\{V_k\}$ contains no information about the function $v(t)$ for values of t outside the interval $(t', t' + T)$.

Thus we have lost something in transforming to the frequency domain $\{V_k\}$ and back to the time domain $v(t)$.

From another viewpoint, consider Fig. 2.5. The function $w(t)$ is identical to the $v(t)$ of Fig. 2.4 in the interval $(t', t' + T)$, but is different outside this interval. If the coefficients $\{W_k\}$ corresponding to $w(t)$ in the interval $(t', t' + T)$ are found, they will be identical to $\{V_k\}$. [Recall that the equations for $v(t)$ and $w(t)$ are identical in the interval $(t', t' + T)$, and this is the function used to calculate the Fourier coefficients in Eq. (2.1).]

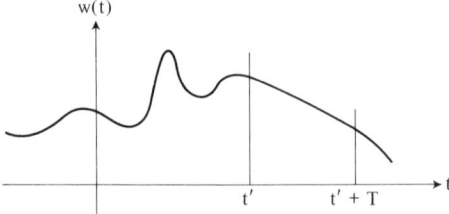

Fig. 2.5 The function $w(t)$ identical to $v(t)$ in the interval $(t', t' + T)$, but different elsewhere.

2.4 FOURIER SERIES EXPANSION OF PERIODIC TIME FUNCTIONS

We conclude that the set of coefficients $\{V_k\}$ is sufficient to specify the time function $v(t)$ in the interval of expansion, but it supplies no information about $v(t)$ outside this interval.

2.4 FOURIER SERIES EXPANSION OF PERIODIC TIME FUNCTIONS

We now consider a periodic time function with period T, again identical to $v(t)$ in the interval $(t', t' + T)$. Since the set of coefficients $\{V_k\}$ completely specifies this time function for one period, the periodicity of the function specifies the function for all time.

Therefore we need to know three things in order to completely specify a periodic time function by its Fourier series coefficients. We need to know

1. that the function is periodic with period T,
2. the period of expansion $(t', t' + T)$,
3. the set of coefficients $\{V_k\}$.

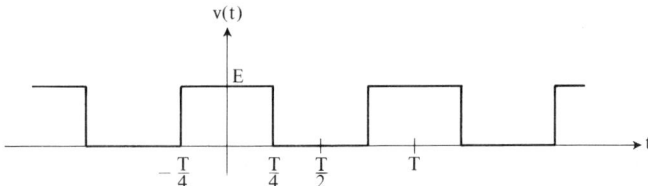

Fig. 2.6 Periodic square wave.

Examples

4. The periodic square wave $v(t)$ is shown in Fig. 2.6. We proceed to find the Fourier coefficients $\{V_k\}$. According to Eq. (2.1), we have

$$V_k = \frac{1}{T}\int_{-T/2}^{T/2} v(t)e^{-j\omega_k t}\,dt = \frac{1}{T}\int_{-T/4}^{T/4} E e^{-j\omega_k t}\,dt$$
$$= E\left(\frac{e^{j\omega_k T/4} - e^{-j\omega_k T/4}}{j\omega_k T}\right). \qquad (2.10)$$

Now $\omega_k = k\omega_1$ and ω_1 is the fundamental frequency related to T by $f_1 = 1/T$. Therefore, $\omega_k T = k\omega_1 T = k2\pi$ and

$$V_k = E\left(\frac{e^{jk2\pi/4} - e^{-jk2\pi/4}}{j2\pi k}\right) = \frac{E}{2}\left(\frac{\sin k\pi/2}{k\pi/2}\right).$$

Figure 2.7 shows a plot of the coefficients V_k versus frequency. The dashed wave is the envelope

$$V(f) = \frac{E}{2}\left(\frac{\sin \omega T/4}{\omega T/4}\right).$$

18   THE FOURIER SERIES

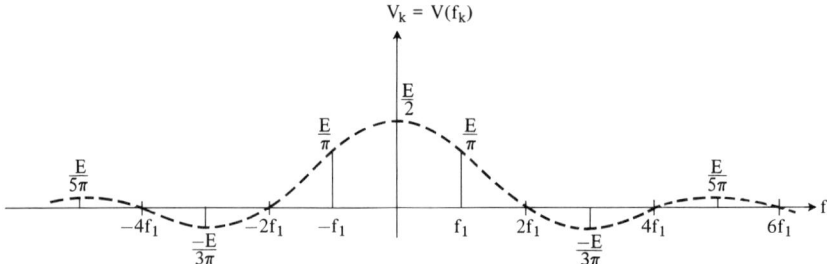

Fig. 2.7 The coefficients V_k versus frequency.

[$V(f)$ is derived from Eq. (2.10) with $\omega = \omega_k$.] The values of the coefficients V_k are given by the values of $V(f)$ evaluated at $V(f) = V(f_k) = V_k$. Thus the coefficients V_k are functions of frequency. Alternatively, we think of the coefficients V_k simply as numbers, and if we know the fundamental period T, we can reconstruct the original periodic time function from these numbers.

Notes. a) Since each of the functions $e^{jk\omega_1 t}$ is periodic, the expansion given in Eq. (2.2) always yields a periodic function with period $T = 2\pi/\omega_1$. The expansion of the functions shown in Figs. 2.4 and 2.5, for example, yields periodic functions with period T.

b) In this example each coefficient is a real (not complex) number. This is a special case. In general, the numbers $\{V_k\}$ are complex. Therefore a graph of the coefficients versus frequency will usually require either a three-dimensional diagram or two separate graphs. In the next example the coefficients are complex numbers.

5. The periodic square wave of Example 4 is shifted to the right by $T/4$ (Fig. 2.8). In this case the coefficients $\{G_k\}$ are given by

$$G_k = \frac{1}{T}\int_0^{T/2} E e^{-j\omega_k t}\, dt = \frac{E}{j\omega_k T}(1 - e^{-j\omega_k T/2})$$

$$= E e^{-j\omega_k T/4}\left(\frac{e^{j\omega_k T/4} - e^{-j\omega_k T/4}}{j\omega_k T}\right),$$

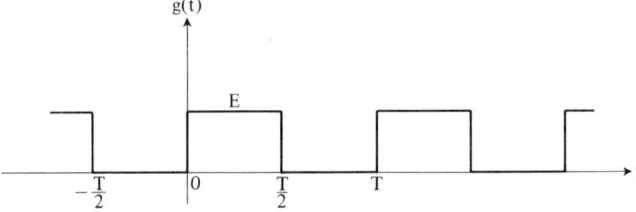

Fig. 2.8 $g(t) = v(t - T/4)$.

2.4 FOURIER SERIES EXPANSION OF PERIODIC TIME FUNCTIONS

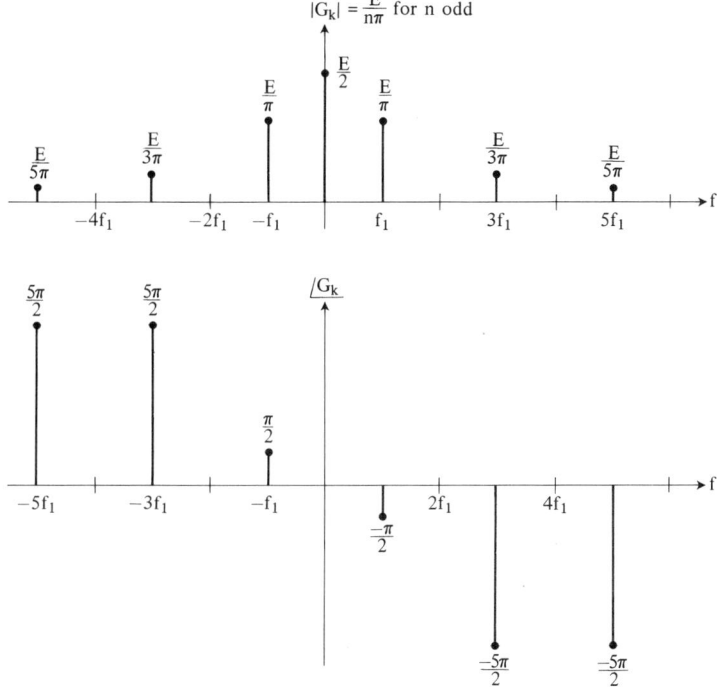

Fig. 2.9 Magnitude and phase of G_k.

which, from Eq. (2.10), is

$$G_k = V_k e^{-j\omega_k T/4} = V_k e^{-jk\pi/2}.$$

A plot of the magnitude and phase of this function of frequency is shown in Fig. 2.9.

Note. Each coefficient G_k is a complex number. This means that G_k is not a single number, but is specified by two numbers (an ordered pair). For instance, G_1 is given by its magnitude and phase as $(E/\pi, -\pi/2)$, which is conventionally written as

$$G_1 = \frac{E}{\pi} e^{-j\pi/2}.$$

The magnitude and phase represent a complex number in polar form. Real and imaginary parts are used to represent this same number in rectangular coordinates. Thus G_1 is given in rectangular coordinates by $(0, -E/\pi)$ or, more conventionally,

$$G_1 = 0 - j\frac{E}{\pi}.$$

(See Fig. 2.10.)

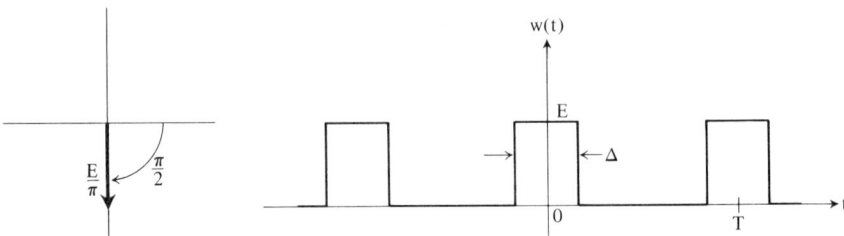

Fig. 2.10 The coefficient G_1.

Fig. 2.11 Periodic square wave with arbitrary pulse width Δ.

6. In Examples 4 and 5 the voltage was on half the period and off half the period. We now examine the more general case of arbitrary pulse width Δ (Fig. 2.11). The coefficients W_k are given by

$$W_k = \frac{1}{T}\int_{-\Delta/2}^{\Delta/2} E e^{-j\omega_k t}\, dt$$

$$= \frac{\Delta E}{T}\left(\frac{\sin k\pi\Delta/T}{k\pi\Delta/T}\right). \tag{2.11}$$

Figure 2.12 shows the graph of W_k.

Notes. a) The spacing between lines is the same as before. This spacing depends on T, the period of the time function. The $(\sin x)/x$ envelope depends on Δ, the pulse width. This $(\sin x)/x$ envelope is characteristic of periodic square pulses. Different pulse shapes result in different envelope shapes.

b) The coefficients are real in Example 6, just as they were in Example 4. This is because the time functions are even. An important property of Fourier series is that even time functions result in purely real coefficients and odd time functions result in purely imaginary coefficients. Other properties are discussed in the next section.

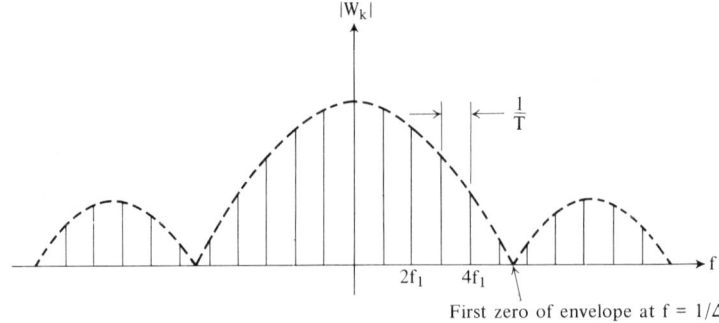

Fig. 2.12 The coefficients W_k.

2.5 PROPERTIES OF FOURIER SERIES

Several useful properties of Fourier series are listed below. The double-headed arrow, $\leftrightarrow$, indicates the one-to-one relationship between the time domain and the frequency domain given by Eqs. (2.1) and (2.2). The arrow is double-headed because of the uniqueness of the Fourier series. That is, if $v(t)$ is transformed into $V(f_k)$ by Eq. (2.1), then $V(f_k)$ is transformed back to the original $v(t)$ by Eq. (2.2).

Recall that $f_k = k f_1$ where $T = 1/f_1$, the fundamental period of the waveform. Thus in Eq. (2.16), $m\omega_1 = m 2\pi f_1$ represents the mth harmonic of the periodic waveform.

Uniqueness: $\quad v(t) \leftrightarrow V(f_k) = V_k \qquad (2.12)$

Superposition: $\quad av(t) + bw(t) \leftrightarrow aV_k + bW_k \qquad (2.13)$

Differentiation: $\quad \dfrac{d^n}{dt^n} v(t) \leftrightarrow (j\omega_k)^n V_k \qquad (2.14)$

Delay: $\quad v(t - t_0) \leftrightarrow V_k e^{-j\omega_k t_0} \qquad (2.15)$

Modulation: $\quad v(t) e^{jm\omega_1 t} \leftrightarrow V_{k-m} \qquad (2.16)$

Convolution: $\quad \dfrac{1}{T} \int_{t'}^{t'+T} v(\lambda) w(t-\lambda) \, d\lambda \leftrightarrow V_k W_k \qquad (2.17)$

Multiplication: $\quad v(t) w(t) \leftrightarrow \sum_{i=-\infty}^{+\infty} V_i W_{k-i} \qquad (2.18)$

Note the duality between

delay—modulation,

convolution—multiplication.

This duality is readily apparent if we write $V(f_k)$ in place of V_k. In Eq. (2.16), V_{k-m} becomes $V(f_k - f_m)$, and in Eq. (2.18), $V_i W_{k-i}$ becomes $V(f_i) W(f_k - f_i)$. This duality is no accident and may be exploited to obtain further interesting properties (see references 2 and 5).

We now prove enough of these properties so that the student should be able to prove the remainder by using similar procedures.

Proof of differentiation. We wish to prove that

$$\dfrac{d^n}{dt^n} v(t) \leftrightarrow (j\omega_k)^n V_k, \qquad (2.14)$$

where $v(t)$ is given by

$$v(t) = \sum_{k=-\infty}^{+\infty} V_k e^{j\omega_k t}. \qquad (2.2)$$

Taking the nth derivative of both sides of Eq. (2.2), we obtain

$$\frac{d^n}{dt^n} v(t) = \sum_{k=-\infty}^{+\infty} (j\omega_k)^n V_k e^{j\omega_k t}.$$

Each coefficient in the series is identified as $(j\omega_k)^n V_k$, and the property is proved.

Proof of delay. We have

$$G_k = \frac{1}{T} \int_{t'}^{t'+T} g(t) e^{-j\omega_k t} \, dt.$$

We identify $g(t)$ as $g(t) = v(t - t_0)$. Then Eq. (2.1) is

$$G_k = \frac{1}{T} \int_{t'}^{t'+T} v(t - t_0) e^{-j\omega_k t} \, dt.$$

We change the variable of integration and let $\lambda = t - t_0$. Then

$$dt = d\lambda, \quad \text{and} \quad \begin{cases} \text{as } t \to t', & \lambda \to t' - t_0, \\ \text{as } t \to t' + T, & \lambda \to t' - t_0 + T. \end{cases}$$

Therefore

$$G_k = \frac{1}{T} \int_{t''}^{t''+T} v(\lambda) e^{-j\omega_k(\lambda + t_0)} \, d\lambda$$

$$= e^{-j\omega_k t_0} \left[\frac{1}{T} \int_{t''}^{t''+T} v(\lambda) e^{-j\omega_k \lambda} \, d\lambda \right]$$

$$= e^{-j\omega_k t_0} V_k,$$

where $t'' = t' - t_0$, and the property is proved.

Proof of convolution. We wish to show that when $g(t)$ is given by

$$g(t) = \frac{1}{T} \int_{t'}^{t'+T} v(\lambda) w(t - \lambda) \, d\lambda, \tag{2.19}$$

the coefficients G_k are given by $V_k W_k$. That is, we wish to show that

$$g(t) = \sum_{k=-\infty}^{+\infty} V_k W_k e^{j\omega_k t}.$$

By Eqs. (2.1) and (2.19), we find that G_k is given by

$$G_k = \frac{1}{T} \int_{t'}^{t'+T} e^{-j\omega_k t} \left[\frac{1}{T} \int_{t'}^{t'+T} v(\lambda) w(t - \lambda) \, d\lambda \right] dt.$$

Changing the order of integration, we get

$$G_k = \frac{1}{T} \int_{t'}^{t'+T} v(\lambda) \left[\frac{1}{T} \int_{t'}^{t'+T} w(t - \lambda) e^{-j\omega_k t} \, dt \right] d\lambda.$$

2.6 EVALUATING COEFFICIENTS BY DIFFERENTIATION PROPERTY

From Eq. (2.15) we recognize the term in brackets as

$$\frac{1}{T}\int_{t'}^{t'+T} w(t - \lambda)e^{-j\omega_k t}\, dt = W_k e^{-j\omega_k \lambda}.$$

Therefore

$$G_k = W_k \left[\frac{1}{T}\int_{t'}^{t'+T} v(\lambda)e^{-j\omega_k \lambda}\, d\lambda\right] = W_k V_k.$$

2.6 USE OF THE DIFFERENTIATION PROPERTY TO EVALUATE COEFFICIENTS

In some cases it is much easier to evaluate the Fourier coefficients indirectly by use of some of the properties. For example, the multiplication and convolution properties can sometimes be used to advantage. Here we introduce the use of the differentiation property, Eq. (2.14), by two examples.

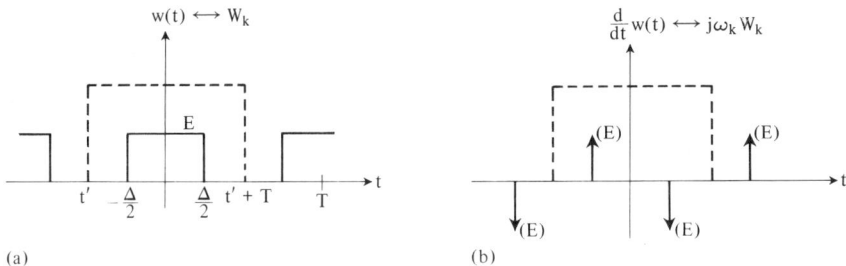

Fig. 2.13

Examples

7. Evaluate the coefficients for $w(t)$ of Example 6. The signal and its derivative are shown in Fig. 2.13 over one period, t' to $t' + T$.

By Eq. (2.1), the coefficients of the waveform shown in Fig. 2.13(b) are given by

$$j\omega_k W_k = \frac{1}{T}\int_{t'}^{t'+T} \frac{d}{dt} w(t)e^{-j\omega_k t}\, dt$$

$$= \frac{1}{T}\int E\delta\left(t + \frac{\Delta}{2}\right)e^{-j\omega_k t}\, dt + \frac{1}{T}\int -E\delta\left(t - \frac{\Delta}{2}\right)e^{-j\omega_k t}$$

$$= \frac{E}{T}(e^{j\omega_k \Delta/2} - e^{-j\omega_k \Delta/2}). \tag{2.20}$$

Solving Eq. (2.20) for W_k, we obtain the same expression as in Eq. (2.11).

For this problem the straightforward method of evaluating the coefficients (Eq. 2.11) was much easier than the method illustrated in this example. We now consider a waveform where this is not so.

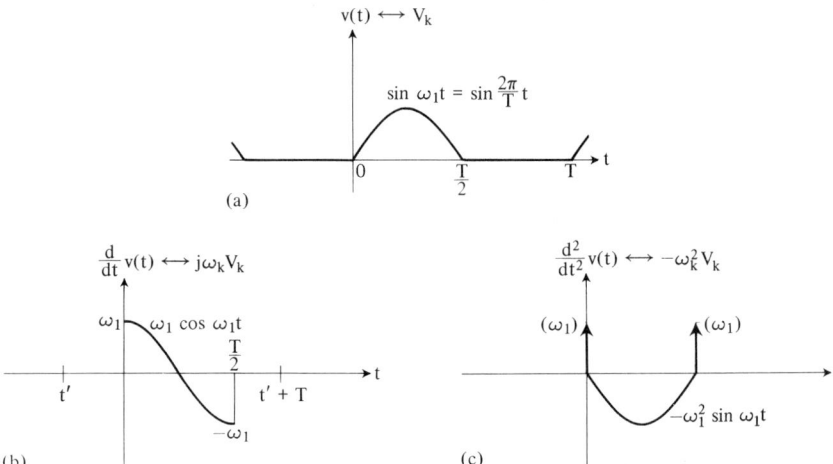

Fig. 2.14

8. The half-wave rectified signal $v(t)$ and its first two derivatives are shown in Fig. 2.14.

Consider Fig. 2.14(c). The waveform would be the same as $v(t)$ (except for the multiplying constant $-\omega_1^2$) if we subtracted the two delta functions. Performing this subtraction, we have

$$\frac{d^2v}{dt^2} - \omega_1 \delta(t) - \omega_1 \delta\left(t - \frac{T}{2}\right) \leftrightarrow -\omega_k^2 V_k - \frac{\omega_1}{T} - \frac{\omega_1}{T} e^{-j\omega_k T/2}. \quad (2.21)$$

But the left-hand side of Eq. (2.21) is proportional to $v(t)$. Therefore

$$\frac{d^2v}{dt^2} - \omega_1 \delta(t) - \omega_1 \delta\left(t - \frac{T}{2}\right) = -\omega_1^2 v(t).$$

The Fourier coefficients for $-\omega_1^2 v(t)$ are given by $-\omega_1^2 V_k$. Therefore from the right-hand side of Eq. (2.21), we have

$$-\omega_1^2 V_k = -\omega_k^2 V_k - \frac{\omega_1}{T} - \frac{\omega_1}{T} e^{-j\omega_k T/2}. \quad (2.22)$$

We now use algebra to solve for V_k from Eq. (2.22), and finally arrive at the solution

$$V_k = \frac{e^{-jk\pi/2} \cos k\pi/2}{\pi(1 - k^2)}.$$

One note of caution should be made. Note that the first derivative of $v(t)$ and the first derivative of $[v(t) + \text{dc-term}]$ are identical. Thus the dc-term V_0 must be evaluated separately when using the differentiation property.

2.7 PARSEVAL'S THEOREM AND POWER SPECTRAL DENSITIES

As we stated in Chapter 1, signals can be classified into two broad categories: power signals and energy signals. If a signal has nonzero power, then it has infinite energy. If a signal has finite energy, then it has zero power (average power). We now wish to discuss some important topics related to the power in periodic signals. In Chapter 3 we will discuss similar topics for energy signals.

Parseval's theorem for periodic signals relates the power in the time domain to the power in the frequency domain. For periodic signals the average power is given by

$$P = \frac{1}{T} \int_{t'}^{t'+T} |v(t)|^2 \, dt = \frac{1}{T} \int_{t'}^{t'+T} v^*(t)v(t) \, dt. \qquad (2.23)$$

The power is related to the Fourier coefficients of $v(t)$. From Eq. (2.2),

$$v(t) = \sum_{k=-\infty}^{+\infty} V_k e^{j\omega_k t}, \qquad (2.2)$$

where $\omega = 2\pi f$, we get

$$P = \frac{1}{T} \int_{t'}^{t'+T} v^*(t) \sum_{k=-\infty}^{+\infty} V_k e^{j\omega_k t} \, dt. \qquad (2.24)$$

By now the reader should realize that in proving anything in which there is a double integration, a double summation, or an integration and a summation, the best thing to try is to interchange the two integrals (or whatever). Thus in Eq. (2.24) we interchange the integration and summation to obtain

$$P = \sum_{k=-\infty}^{+\infty} V_k \left[\frac{1}{T} \int_{t'}^{t'+T} v^*(t) e^{j\omega_k t} \, dt \right].$$

The term in brackets is the complex conjugate of V_k. This leads to Eq. (2.25), which is known as *Parseval's theorem:*

$$P = \frac{1}{T} \int_{t'}^{t'+T} |v(t)|^2 \, dt = \sum_{k=-\infty}^{+\infty} |V_k|^2. \qquad (2.25)$$

If we know the time function $v(t)$, we can find the power. Alternatively, if we know the frequency function $V_k = V(f_k)$, we can find the power.

Example 9. For the sinusoid,

$$v(t) = \sin 377t. \qquad (2.26)$$

The average power, from Eq. (2.23), is

$$P = \frac{1}{T} \int_0^T \sin^2 377t = \tfrac{1}{2}.$$

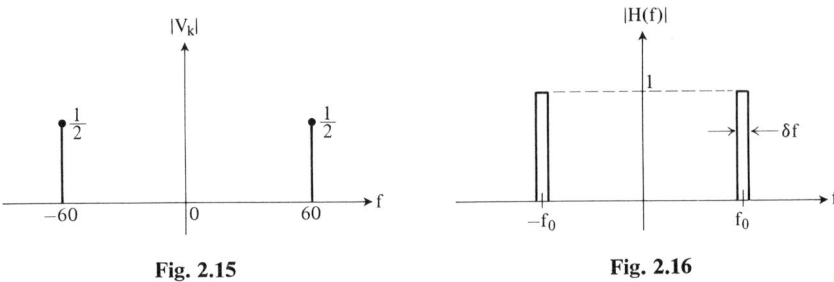

Fig. 2.15 Fig. 2.16

The Fourier coefficients are given by

$$V_{-1} = \tfrac{1}{2}e^{-j\pi/2},$$
$$V_1 = \tfrac{1}{2}e^{j\pi/2},$$
$$V_k = 0 \quad \text{for all other } k.$$

(See Fig. 2.15.) Using the right-hand side of Eq. (2.25), we obtain

$$P = \sum_{k=-\infty}^{+\infty} |V_k|^2 = \tfrac{1}{4} + \tfrac{1}{4} = \tfrac{1}{2},$$

which is the same as the answer obtained in the time domain, as it should be.

Power Spectral Density

Suppose an experiment is performed in the laboratory using the narrow band filter shown in Fig. 2.16. Only the magnitude of the transfer function is shown in the figure. The filter passes all frequency components in the narrow band of δf centered about f_0 and rejects all others. Suppose that the filter is tunable, so that we may vary the center frequency f_0. (An approximate example of such a filter is the *RLC* tank circuit used to select radio stations in an ordinary AM radio.)

A wattmeter is connected to the output of the filter. As the center frequency f_0 is varied throughout the frequency range, we would expect the wattmeter reading to change. For a sinusoidal input, say $v(t)$ of Eq. (2.26), the wattmeter would read zero except when f_0 was close to 60 cps. In the neighborhood of 60 cps, the wattmeter would read $\tfrac{1}{2}$ W (watt).

Suppose the input to the filter is a periodic square wave (Example 6). Then the wattmeter reading will be zero except when f_0 is close to f_1, $2f_1$, $3f_1$, etc. At these frequencies the wattmeter reading will be $2V_1^2$, $2V_2^2$, $2V_3^2$, etc., by Parseval's theorem. The V_i^2-terms are doubled because the filter is tuned to $+f_0$ and $-f_0$, thus passing components at both frequencies. Of course, there is no such thing as negative frequency in the laboratory, but the mathematics is greatly simplified if the double-sided spectrum is used.

As a third example, suppose the input to the filter is a general power signal, not necessarily periodic. The wattmeter reading will vary as the

2.7 PARSEVAL'S THEOREM AND POWER SPECTRAL DENSITIES

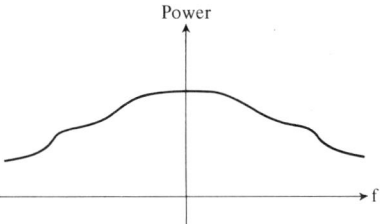

Fig. 2.17 Wattmeter reading versus frequency.

frequency f_0 is tuned from zero to infinity. A plot of wattmeter reading versus frequency such as Fig. 2.17 might result. This plot is made by dividing the wattmeter reading by two and representing one-half of the power at positive frequencies and one-half at negative frequencies. This curve contains information about the distribution of power along the frequency axis.

It would be nice if, in addition to specifying the location of power, the curve indicated the amount of power in each frequency band. Such a curve is called the *power spectral density*, $G(f)$. Then the total power in the frequency band $f_1 < f < f_2$ is given by

$$P_{f_1 < f < f_2} = \int_{f_1}^{f_2} G(f)\,df. \tag{2.27}$$

Later, in our study of random power signals, we will derive the power spectral density for general signals. For the time being, we are interested in the relationship between the Fourier coefficients V_k and the power spectral density. The power at any harmonic frequency f_k is just V_k^2. Since we wish to use integrals, we will represent the power at the harmonic frequencies by delta functions. Therefore, for periodic signals the power spectral density $G(f)$ is given by a series of delta functions, and the areas under these delta functions are given by V_k^2.

Example 10. The periodic square wave of Example 4 has the power spectral density shown in Fig. 2.18. The total power in the range $|f| < 4f_1$ is approxi-

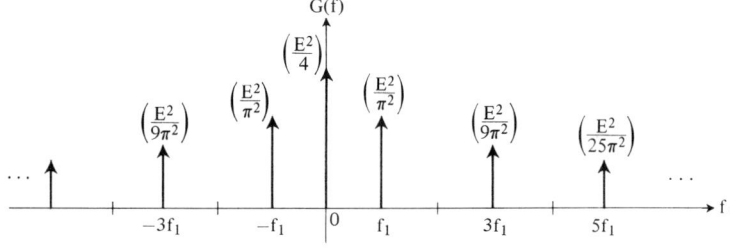

Fig. 2.18

mately $0.475E^2$, and is given by

$$P_{|f|<4f_1} = \sum_{k=-3}^{+3} |V_k|^2 \approx 0.475E^2.$$

Note that this is 95 percent of the total power. The total power is found in the time domain to be

$$P_{\text{total}} = \frac{1}{T}\int_0^T v^2(t)\,dt = 0.5E^2.$$

2.8 EXAMPLES

In this section we provide some examples of the use of the theory given in Section 2.7 in applications to circuits. Also, we introduce here some concepts that the student should already be familiar with but may need some review of.

11. Find the frequency spectrum and the power density spectrum for the periodic triangular wave shown in Fig. 2.19. The coefficients are given by

$$V_k = \frac{1}{T}\int_{t'}^{t'+T} v(t)e^{-j2\pi f_k t}\,dt, \qquad \text{where} \qquad f_k = kf_1 = \frac{k}{T},$$

$$V_k = \frac{1}{T}\int_0^T \frac{E}{T} t\, e^{-j2\pi f_k t}\,dt = \begin{cases} \dfrac{jE}{2k\pi}, & k \neq 0, \\ \dfrac{E}{2}, & k = 0. \end{cases}$$

Or, calculating magnitude and phase, the coefficients are given in the form of $V_k = |V_k|e^{j\theta_k}$ as follows:

$$V_k = \begin{cases} \dfrac{E}{2k\pi}e^{j\pi/2}, & k > 0, \\ \left|\dfrac{E}{2k\pi}\right|e^{-j\pi/2}, & k < 0, \\ \dfrac{E}{2}, & k = 0. \end{cases}$$

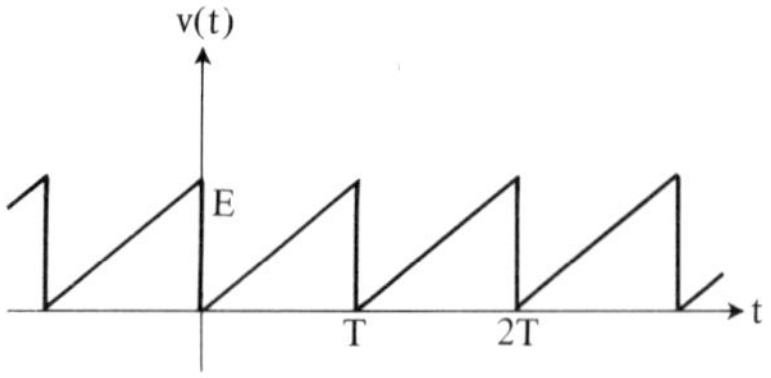

Fig. 2.19

The plot of this function, in magnitude and phase, is shown in Fig. 2.20.

The power spectral density is found by taking each V_k-term and squaring it. The plot of this function is shown in Fig. 2.21.

12. The periodic sawtooth waveform of Example 11 has period $T = 10^{-3}$ sec. This waveform is applied to an ideal low-pass filter with zero phase shift and cutoff frequency $f_c = 1500$ cps. Find the output voltage as a function of time and the rms (root mean square) value of the output.

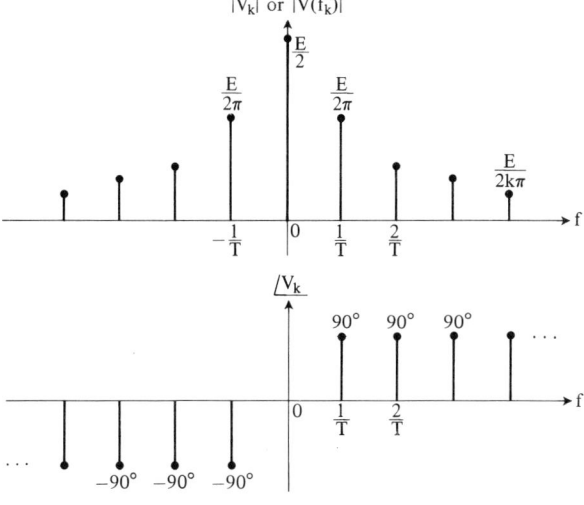

Fig. 2.20

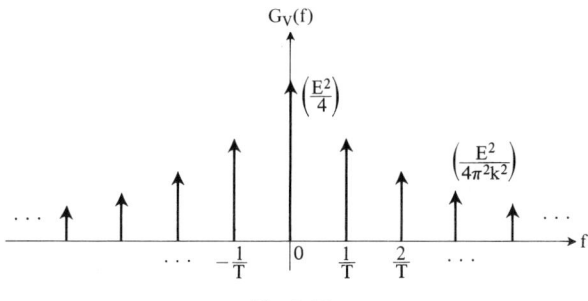

Fig. 2.21

The filter transfer function $H(f)$ is shown in Fig. 2.22; the components of V_k that are inside the cutoff frequency are shown as dashed lines. The filter passes V_{-1}, V_0, and V_1. Call the filter output $y(t)$. Then

$$y(t) = \sum_{k=-\infty}^{+\infty} V_k H(f) e^{j\omega_k t} = \sum_{k=-1}^{+1} V_k e^{j\omega_k t}$$

$$= \frac{E}{2\pi} e^{-j\pi/2} e^{-j\omega_1 t} + \frac{E}{2} + \frac{E}{2\pi} e^{j\pi/2} e^{j\omega_1 t}$$

$$= \frac{E}{2} + \frac{E}{\pi} \cos\left(\omega_1 t + \frac{\pi}{2}\right)$$

$$= \frac{E}{2} - \frac{E}{\pi} \sin(\omega_1 t), \qquad (2.28)$$

where $\omega_1 = 2\pi/T = 2\pi(1000)$ rad/sec.

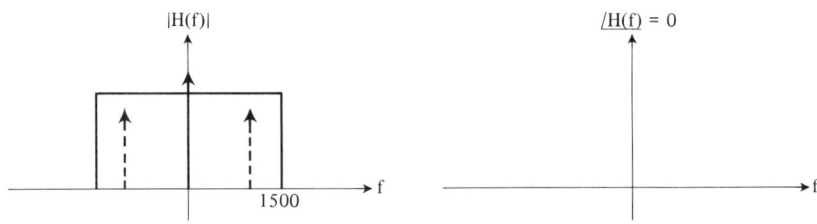

Fig. 2.22

The rms value of $y(t)$ may be found from the power spectral density, Fig. 2.21, or may be calculated directly from Eq. (2.28). Since the components V_{-1}, V_0, and V_1 are unaltered, we have, from Fig. 2.21, that the mean-square value of $y(t)$ is

$$Y_{\text{ms}} = \sum_{k=-1}^{+1} G_V(f) = \frac{E^2}{4} + \frac{E^2}{2\pi^2}$$

and the rms value is the square root of this.

Note. Euler's formula is used to relate the exponential and sinusoidal forms of Eq. (2.28). Euler's formula is

$$e^{\pm j\theta} = \cos \theta \pm j \sin \theta$$

or, equivalently,

$$\cos \theta = \frac{e^{j\theta} + e^{-j\theta}}{2},$$

$$\sin \theta = \frac{e^{j\theta} - e^{-j\theta}}{2j}.$$

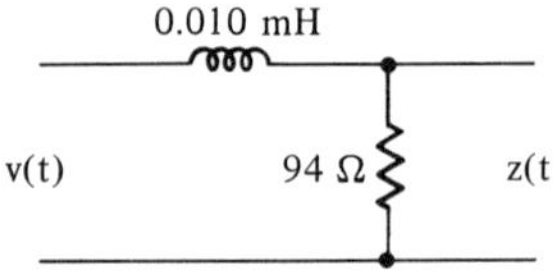

Fig. 2.23

13. Suppose the low-pass RL-filter of Fig. 2.23 is used as an approximation to the ideal filter of Fig. 2.22. The filter characteristic is shown in Fig. 2.24. The value of R and L are adjusted so that the 3-db filter cutoff is 1500 cps. Now calculate the rms value of the output.

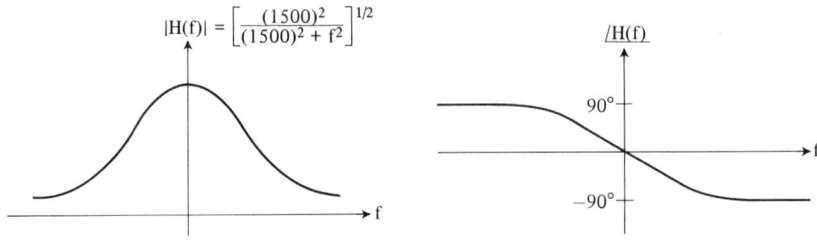

Fig. 2.24 The filter characteristic.

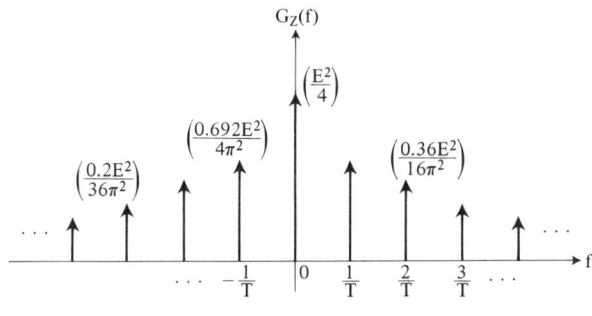

Fig. 2.25

Call the *RL*-filter output $z(t)$. The filter output at a particular frequency f_k is given by

$$Z(f_k) = H(f)V(f_k).$$

Therefore, the square of Z at any frequency f_k is found by squaring $Z(f_k)$. Thus we get

$$|Z(f_k)|^2 = |H(f)|^2 |V(f_k)|^2.$$

This allows us to find the power spectral density of the output, $G_Z(f)$, from

$$G_Z(f) = G_V(f) |H(f)|^2. \quad (2.29)$$

This function is plotted in Fig. 2.25.

The total power (on a one-ohm basis) at the filter output is the mean-square value:

$$Z_{\text{ms}} = \sum_{k=-\infty}^{+\infty} G_Z(f_k) \approx 0.29E^2.$$

(Only the first few terms contribute significantly to the sum.) The rms value is the square root of this:

$$Z_{\text{rms}} \approx 0.54E.$$

14. In this example we will find the Fourier series for the product of sinusoidal waveforms. A product device is shown in Fig. 2.26; the input is two cosine waveforms. Solving for the product, we have

$$v(t) = \cos 2\pi t \cdot \cos 2\pi(2)t$$

$$= \left(\frac{e^{j2\pi t} + e^{-j2\pi t}}{2}\right)\left(\frac{e^{j4\pi t} + e^{-j4\pi t}}{2}\right)$$

$$= \frac{e^{j6\pi t} + e^{-j6\pi t} + e^{j2\pi t} + e^{-j2\pi t}}{4}$$

$$= \tfrac{1}{2} \cos 2\pi(3)t + \tfrac{1}{2} \cos 2\pi t.$$

The spectrum of the output $V(f_k)$ is plotted in Fig. 2.27.

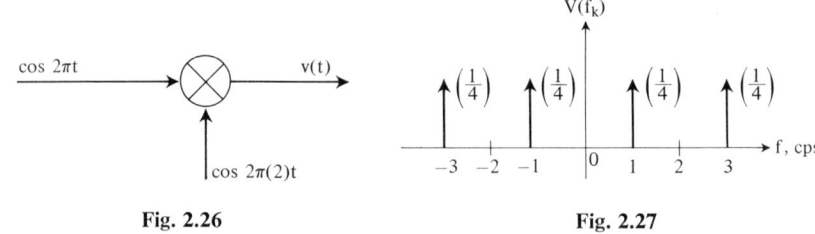

Fig. 2.26   **Fig. 2.27**

Notes. a) This same result may be derived by use of either the modulation property, Eq. (2.16), or the multiplication property, Eq. (2.18).

b) Only the magnitude spectrum is plotted in Fig. 2.27. The phase spectrum is zero.

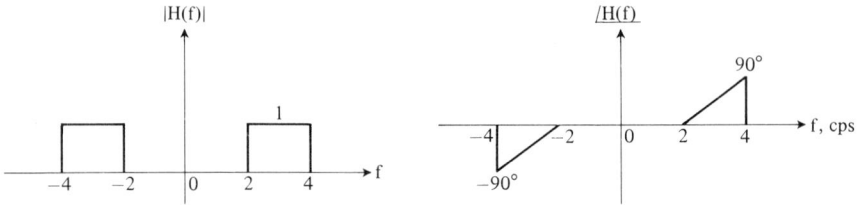

Fig. 2.28 Magnitude and phase for an ideal bandpass filter.

15. We now apply the signal $v(t)$ in Example 14 to the bandpass filter shown in Fig. 2.28. We wish to find the filter output $y(t)$.

The output Y in the frequency domain is given by

$$Y(f) = H(f)V(f), \tag{2.30}$$

where

$$H(f) = |H(f)|\, e^{j\angle H(f)} \tag{2.31}$$

and

$$V(f) = |V(f)|\, e^{j\angle V(f)}, \tag{2.32}$$

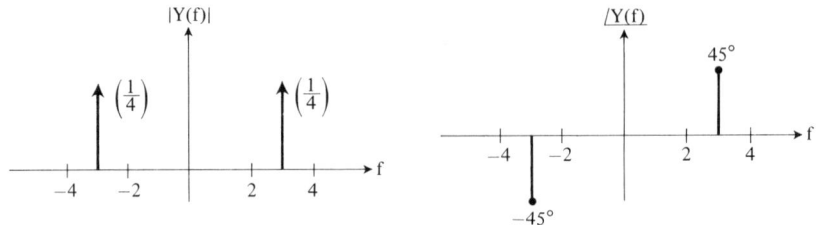

Fig. 2.29

where the input $V(f)$ has values only at $f = f_k$. Combining Eqs. (2.30), (2.31), and (2.32), we have

$$Y(f) = |Y(f)| e^{j\underline{/Y(f)}} = |H(f)| |V(f)| e^{j(\underline{/H(f)} + \underline{/V(f)})}.$$

This output is plotted in Fig. 2.29. From Eq. (2.2) we find that the time function $y(t)$ is

$$y(t) = \tfrac{1}{4} e^{j45°} e^{j\omega_1 t} + \tfrac{1}{4} e^{-j45°} e^{-j\omega_1 t},$$

where

$$\omega_1 = 2\pi(3) \text{ rad/sec},$$
$$y(t) = \tfrac{1}{4} e^{j(\omega_1 t + 45°)} + \tfrac{1}{4} e^{-j(\omega_1 t + 45°)}$$
$$= \tfrac{1}{2} \cos(2\pi 3 t + 45°).$$

PROBLEMS

1. Show that ordinary signals generated in the laboratory are vectors.
2. Find and plot the complex exponential Fourier series for
   a) $f(t) = 50 \sin 2\pi(100)t$,
   b) $g(t) = 50 \cos 2\pi(100)t$.
3. The Fourier coefficients V_k are given by

$$V_k = 1, \quad k = \pm 2, \pm 1, 0,$$
$$V_k = 0 \quad \text{for all other } k.$$

a) Evaluate $v(t)$ at $t = 0$.
b) Evaluate $v(t)$ at $t = 1$.
c) Can you write an equation for $v(t)$ for all t?
d) What is the fundamental period of $v(t)$?

4. The amplitude and phase spectrum of the periodic signal $v(t)$ are shown in Fig. 2.30. Write an analytical expression for $v(t)$ in terms of trigonometric functions.

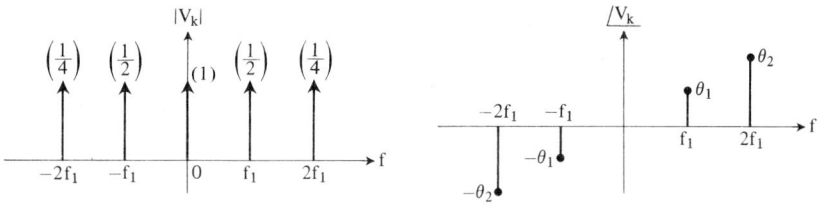

Fig. 2.30

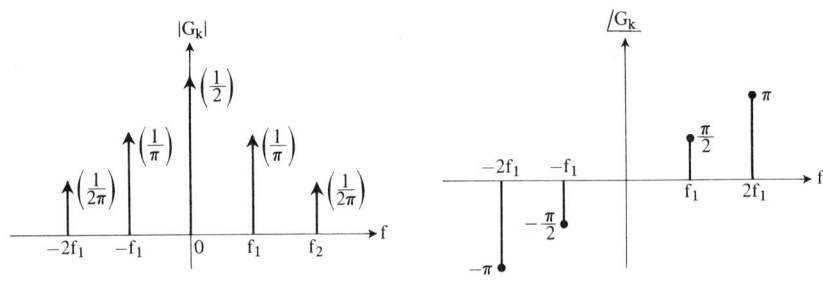

Fig. 2.31

5. Find and plot the time function $g(t)$ that is represented by the Fourier series shown in Fig. 2.31.

6. Prove all properties given for Fourier series (except uniqueness).

7. The signal $f(t)$ shown in Fig. 2.32 is supplied to an ideal low-pass filter with cutoff frequency 20 cps. Find the rms value of the output $y(t)$.

8. A periodic time function $f(t)$ in volts is shown in Fig. 2.33. For $a = \frac{1}{2}$ this time function is a unit square wave.

a) For $a = \frac{1}{2}$ compute the magnitude of the voltage at each harmonic and make a sketch of the spectrum of $f(t)$.

b) Determine the effect of a very small departure of $f(t)$ from a perfect square wave. That is, let $a = \frac{1}{2}(1 + \Delta)$, where Δ is small compared to 1 (but is not 0).

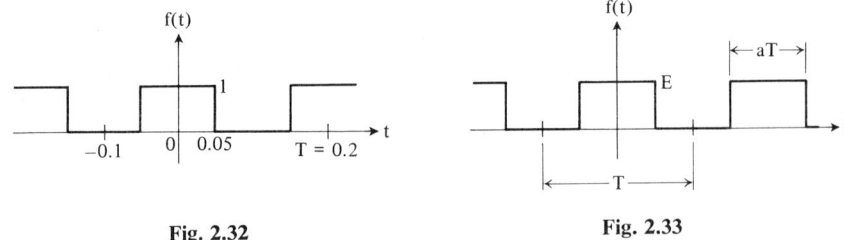

Fig. 2.32 Fig. 2.33

9. The waveform of Problem 8 is passed through a narrow band filter with unity voltage gain. When the filter is tuned to pass the fundamental, the measured output is 1 V (volt) rms. When the filter is tuned to the second harmonic, an output of $\frac{1}{10}$ V rms is measured.

a) Find the amplitude E and the value of a in the input waveform.
b) Find the total rms value of the input waveform.

10. A square wave of unit amplitude and 1-kcs fundamental frequency is passed through an ideal low-pass filter with unity voltage gain and 5.3-kcs bandwidth. (See Fig. 2.34.)

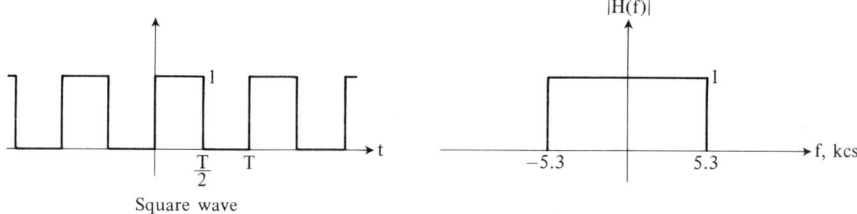

Fig. 2.34

a) Find the output rms voltage and the ratio of input to output power on a one-ohm basis.
b) What is the output power on a one-ohm basis?
c) What is the output power if the load resistor is 1 kΩ?

11. The period of the square wave $A(t)$ is 100 μsec. This voltage is applied to an ideal low-pass filter with 25-kcs bandwidth and gain of 1. (See Fig. 2.35.) What is the total rms voltage at the output of the filter?

12. An approximation to the ideal low-pass filter is the simple circuit shown in Fig. 2.36.

a) With $R = 10$ kΩ, find the value of C so that this filter has the same bandwidth as the ideal filter of Problem 11. Use 3-db bandwidth.
b) Assume the same input as Problem 11 and compute the total rms output.

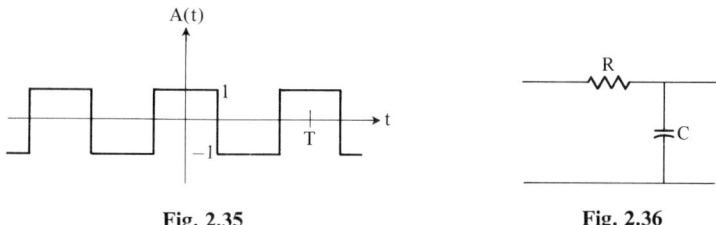

Fig. 2.35 **Fig. 2.36**

13. The period of $A(t)$ in Fig. 2.35 is 10 μsec for this problem. Consider the following two waveforms:

$$e_1(t) = A(t)\cos \omega_0 t, \quad e_2(t) = \cos \omega_0 t,$$

where

$$\omega_0 = 2\pi(10)^7.$$

In the time domain, $e_1(t)$ and $e_2(t)$ look remarkably alike. Yet if these waveforms are passed through a filter with 50-kc bandpass centered at 10 mc, the outputs are entirely different. Find the steady-state output voltage, $e_0(t)$, for both input waveforms (assume the gain of filter is 1 and the phase is 0). (See Fig. 2.37.)

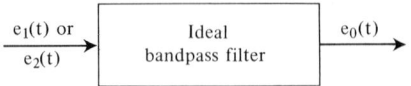

Fig. 2.37

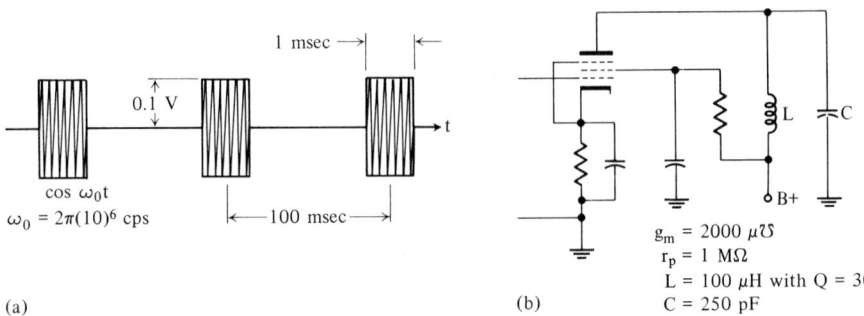

Fig. 2.38

14. The periodic RF pulse waveform shown in Fig. 2.38 is applied to the grid circuit of a pentode amplifier. Assume that the tube is in class-A operation. Compute the rms output voltage at the plate of the tube.

15. Sketch and carefully label the magnitude and phase of the spectrum of each of the following, where $\omega_A = 1$ kcs, $\omega_0 = 5$ kcs, and $m = 1$.
  a) $f(t) = \cos \omega_0 t$
  b) $f(t) = \sin \omega_0 t$
  c) $f(t) = \cos \omega_A t \cos \omega_0 t$
  d) $f(t) = (1 + m \cos \omega_A t) \cos \omega_0 t$
  e) $f(t) = \cos (\omega_0 t + \phi)$
  f) $f(t) = \cos^2 (\omega_0 t)$

16. Find power (mean-square value) for each waveform in Problem 15.

17. One proof of Parseval's theorem (Eq. 2.25) is given in the text. An alternative proof uses the fact that the exponential functions

$$\phi_k = e^{jk2\pi t/T} = e^{j\omega_k t}$$

are orthogonal over any interval of length T. Construct this proof. [*Hint:* Begin with the definition of inner product divided by T,

$$\frac{\langle v(t), v(t) \rangle}{T} = \frac{1}{T} \int_{t'}^{t'+T} |v(t)|^2 \, dt,$$

and make use of Eq. (2.2) to express $v(t)$ in terms of V_k.]

18. Referring to Fig. 2.39,
  a) find and sketch the magnitude of the frequency spectra of $f(t), g(t)$, and $h(t)$;
  b) find and plot the power density spectrum of $f(t)$ as a function of frequency.

19. Given $v(t) = 2 \cos (2\pi 100 t) + 4 \cos (2\pi 100 t) \cos(2\pi 50 t)$, find $y(t)$ in Fig. 2.40.

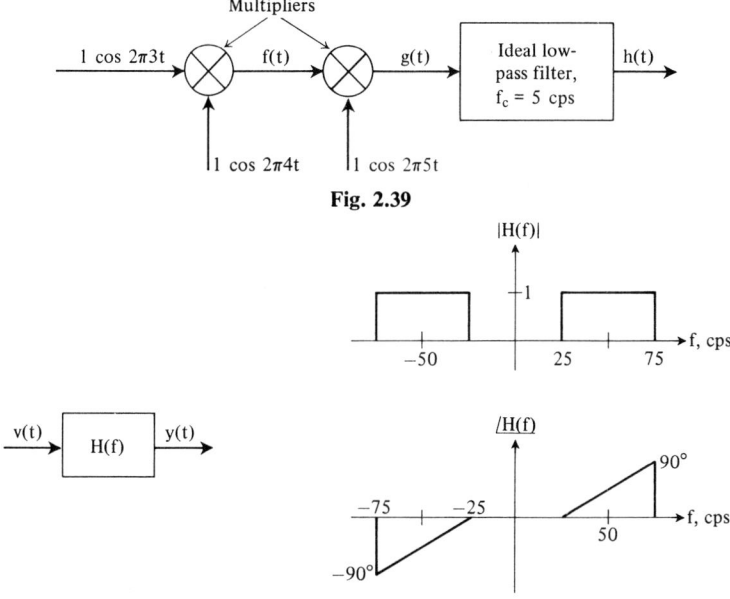

Fig. 2.39

Fig. 2.40

20. The waveform shown in Fig. 2.41 consists of a dc-voltage plus a single pulse. A waveform $f_2(t)$ is formed by multiplying $f_1(t)$ by $\cos \omega_0(t)$. Thus we have

$$f_2(t) = f_1(t) \cos \omega_0 t, \qquad \omega_0 = 2\pi(10)^6 \text{ cps}.$$

What is the rms voltage of the waveform $f_2(t)$?

21. The input to a half-wave rectifier (Fig. 2.42) is given by

$$v_1(t) = V_0 \cos \omega_0 t + V_1 \cos (\omega_0 + \omega_A)t, \qquad \omega_A \ll \omega_0.$$

Assume that $V_1 \ll V_0$ and find

a) the spectrum of the output,
b) the output time function if the half-wave rectifier is followed by a low-pass filter with bandwidth $2\omega_A$.

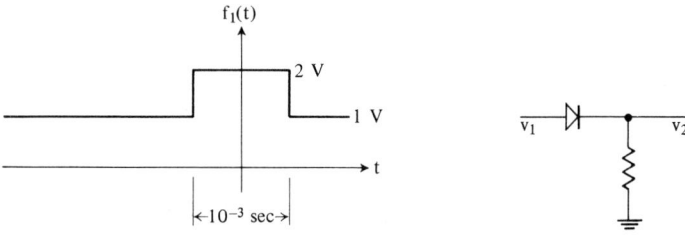

Fig. 2.41　　　　　　　　　　Fig. 2.42

38 THE FOURIER SERIES

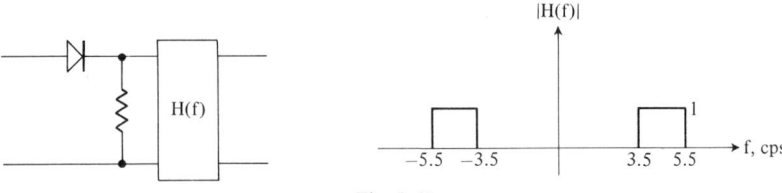

Fig. 2.43

22. The signal $v(t) = \cos 2\pi t$ is supplied to the diode circuit followed by the ideal bandpass filter shown in Fig. 2.43. Find the rms value of the filter output on a one-ohm basis.

FURTHER READING

1. M. McFadden, *Sets, Relations, and Functions*, McGraw-Hill, New York, 1963.
2. W. Kaplan, *Operational Methods for Linear Systems*, Addison-Wesley, Reading, Mass., 1962.
3. A. Papoulis, *The Fourier Integral*, McGraw-Hill, New York, 1962.
4. B. P. Lathi, *Signals, Systems, and Communications*, Wiley, New York, 1965.
5. S. J. Mason, and H. J. Zimmerman, *Electronic Circuits, Signals, and Systems*, Wiley, New York, 1960.

If the reader is unfamiliar with the concept of function, the programed text by McFadden offers a thorough introduction. The text by Kaplan discusses the concept of operator and its application to system theory in Chapter 2. References 3, 4, and 5 provide general discussions of Fourier series.

3
THE FOURIER TRANSFORM

3.1 THE FOURIER TRANSFORM AS AN OPERATOR

The Fourier transform is an operator. It is given by

$$V(f) = \int_{-\infty}^{+\infty} v(t)e^{-j2\pi ft}\, dt. \tag{3.1}$$

Again, as in Section 2.1, this operator is as pictured in Fig. 3.1. One of the elements in the domain [a particular function $v(t)$] is selected and fed into the function machine "Fourier transform": what comes out is the corresponding function $V(f)$.

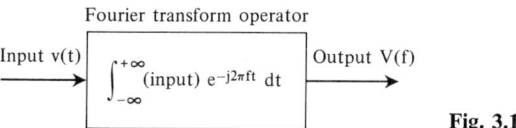

Fig. 3.1

Notes. a) v and V are themselves functions. In our application, $v(t)$ will be identified as a function of time and $V(f)$ will be a function of frequency. That is, the domain of v is a set of numbers that represent values of time, and the domain of V is a set of numbers that represent values of frequency.

b) The parameter f is continuous, in contrast to the parameter f_k for the Fourier series, which is discrete.

c) The range of both v and V can be a set of complex numbers. Since we will be most often concerned with signals that can be generated in the laboratory, the functions v will, for the most part, be real-valued.

The inverse operator is given by

$$v(t) = \int_{-\infty}^{+\infty} V(f)e^{j2\pi ft}\, df \tag{3.2}$$

40 THE FOURIER TRANSFORM 3.1

or, equivalently,

$$v(t) = \int_{-\infty}^{+\infty} V(\omega) e^{j\omega t} \frac{d\omega}{2\pi}. \tag{3.3}$$

Equation (3.3) is related to Eq. (3.2) by a simple change of variable, $\omega = 2\pi f$.

Note. As a general rule of thumb, whenever the variable of integration is ω, divide the integral by 2π. Whenever the variable of integration is f, don't divide by 2π. As with all such rules of thumb, this must be used with caution; but at least one should guard against leaving off the 2π when integrating with respect to ω. This applies to all integrals, not just the Fourier transform. For example, Eq. (2.27) is

$$P_{f_1 < f < f_2} = \int_{f_1}^{f_2} G(f)\, df. \tag{2.27}$$

By the change of variable $\omega = 2\pi f$,

$$P_{\omega_1 < \omega < \omega_2} = \int_{\omega_1}^{\omega_2} G(\omega) \frac{d\omega}{2\pi}. \tag{3.4}$$

The inverse Fourier transform is pictured in Fig. 3.2. The input to the function machine is selected from a set of frequency functions, and the corresponding output is a member of a set of time functions.

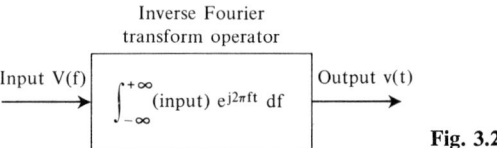

Fig. 3.2

Specifying the Domain

Notice the similarity between the forward and inverse operators, Eqs. (3.1) and (3.2). The only difference is the sign of the exponent. Therefore the conditions that specify the domain of the forward operator are precisely those conditions that specify the domain of the inverse operator; that is,

$$\int_{-\infty}^{+\infty} |v(t)|\, dt < \infty, \tag{3.5}$$

$$\int_{-\infty}^{+\infty} |v(t)|^2\, dt < \infty. \tag{3.6}$$

For $v(t)$ to have a Fourier transform given by Eq. (3.1), it is sufficient that Eqs. (3.5) and (3.6) be satisfied. If Eq. (3.6) is satisfied but Eq. (3.5) is not, then $v(t)$ has a Fourier transform, but it may not be given by Eq. (3.1).

Notes. a) If a function v satisfies Eqs. (3.5) and (3.6), then the Fourier transform $V(f)$ exists and

$$\left[\int_{-\infty}^{+\infty} |V(f)|^2 \, df\right]^{1/2} = \left[\int_{-\infty}^{+\infty} |v(t)|^2 \, dt\right]^{1/2}.$$

b) Any function $V(f)$ that satisfies Eqs. (3.5) and (3.6) [with $V(f)$ substituted for $v(t)$ and integrated with respect to the variable f] has a Fourier transform $v(t)$ given by Eq. (3.2).

c) As in Chapter 2, the criteria used to specify the domain are not both necessary and sufficient. In fact, this problem has not been solved. Many methods of specifying sufficient conditions on v are available, but no one has found necessary conditions for v to have a Fourier transform.

d) Equation (3.6) is precisely our definition of energy signals. This is, of course, the reason for specifying the domain in this manner.

We now have methods to describe energy signals and periodic power signals in the frequency domain. As yet we have no such methods to describe nonperiodic power signals. We will encounter this problem later.

3.2 RELATIONSHIP BETWEEN THE FOURIER TRANSFORM AND OTHER OPERATORS

Relationship Between the Fourier Transform and the Fourier Series

The Fourier series and the Fourier integral are both operators. These operators map time functions into frequency functions, and the inverse operators map frequency functions into time functions. Let us consider the frequency functions.

1. The Fourier series. The domain is discrete. The function $V(f_k)$ has values (or is defined) only for discrete frequencies f_k, $k = \ldots, -2, -1, 0, 1, 2, 3, \ldots$. The range is complex. Values in the range are characterized by two numbers, either real and imaginary parts or magnitude and phase.

2. The Fourier transform. The domain is continuous. The function $V(f)$ has values (or is defined) for all frequencies. The range is complex.

Thus, we see that the only difference in the frequency functions is in the nature of the domain (discrete or continuous).

Let us now consider the time functions.

1. The Fourier series. The orthogonal expansion is defined for a finite time interval $(t', t' + T)$. Thus the series is useful for periodic power signals, but there is no reason to restrict ourselves to periodic signals. If we are interested only in the interval $(t', t' + T)$, it makes no difference whether we are dealing with power signals, energy signals, periodic signals, or whatnot.

2. *The Fourier transform.* The integral is taken over all time $(-\infty, +\infty)$, and thus the time function should be an energy signal. At least the criterion we are using to specify the domain of the forward operator restricts the Fourier transform to energy signals.

Relationship Between the Fourier and Laplace Transforms

The Fourier and Laplace transforms are related in two ways: by the fact that they are both operators, and by the convergence factor σ.

Let us first define $F(R)$ as

$$F(R) = \int_{-\infty}^{+\infty} f(t)e^{-Rt}\, dt, \tag{3.7}$$

which holds for all $f(t)$ in the domain. Note that F is a function of whatever t is multiplied by in the exponent. The Fourier transform pair is

$$F(j\omega) = \int_{-\infty}^{+\infty} f(t)e^{-j\omega t}\, dt, \tag{3.8}$$

$$f(t) = \int_{-\infty}^{+\infty} F(j\omega)e^{j\omega t}\frac{d\omega}{2\pi}. \tag{3.9}$$

We now proceed to derive the Laplace transform pair. Let $v(t) = f(t)e^{-\sigma t}$. Then Eq. (3.8) gives

$$V(j\omega) = \int_{-\infty}^{+\infty} v(t)e^{-j\omega t}\, dt = \int_{-\infty}^{+\infty} f(t)e^{-(\sigma+j\omega)t}\, dt = F(\sigma + j\omega) \tag{3.10}$$

by Eq. (3.7). With $s = \sigma + j\omega$, this yields the forward transform $F(s)$. Therefore

$$\mathcal{F}[f(t)e^{-\sigma t}] = \mathcal{L}[f(t)].$$

where $\mathcal{F}[\ ]$ denotes the Fourier transform and $\mathcal{L}[\ ]$ denotes the two-sided Laplace transform.

You may be used to seeing a lower limit of zero instead of minus infinity in Eq. (3.10). The one-sided Laplace transform commonly used in introductory circuit theory has limits $(0, \infty)$. The two-sided transform has limits $(-\infty, +\infty)$. The two transforms are identical if $f(t) = 0$ for $t < 0$.

We now wish to find the inverse Fourier transform of $F(\sigma + j\omega)$. But $F(\sigma + j\omega) = V(j\omega)$, and the inverse $\mathcal{F}$-transform of $V(j\omega)$ is $v(t)$. Therefore

$$v(t) = e^{-\sigma t}f(t) = \int_{-\infty}^{+\infty} F(\sigma + j\omega)e^{j\omega t}\frac{d\omega}{2\pi}$$

or

$$f(t) = \int_{-\infty}^{+\infty} F(\sigma + j\omega)e^{(\sigma+j\omega)t}\frac{d\omega}{2\pi}.$$

3.2 THE FOURIER TRANSFORM AND OTHER OPERATORS

We now change the variable of integration and let $s = \sigma + j\omega$. Then

$$d\omega = \frac{ds}{j}, \quad \text{and} \quad \begin{cases} \text{as } \omega \to +\infty, & s \to \sigma + j\infty, \\ \text{as } \omega \to -\infty, & s \to \sigma - j\infty. \end{cases}$$

Therefore

$$f(t) = \int_{\sigma-j\infty}^{\sigma+j\infty} F(s)e^{st} \frac{ds}{2\pi j}.$$

We can thus state the relationship between the inverse transforms as

$$\mathscr{F}^{-1}[F(\sigma + j\omega)e^{\sigma t}] = \mathscr{L}^{-1}[F(s)].$$

The convergence factor σ is built into the $\mathscr{L}$-transform. For example, the $\mathscr{F}$-transform of a unit step $u(t)$ does not exist in the classical sense (although it does exist as distribution). However, we can multiply the step by $e^{-\sigma t}$, where σ is any positive number. Then the $\mathscr{F}$-transform of $u(t)e^{-\sigma t}$ does exist and is given by $U(\sigma + j\omega)$.

The inverse $\mathscr{F}$-transform of $U(\sigma + j\omega)$ does not equal $u(t)$. However, $\mathscr{F}^{-1}[U(\sigma + j\omega)e^{\sigma t}]$ does equal $u(t)$. But what we have done is simply use $\mathscr{L}$-transforms.

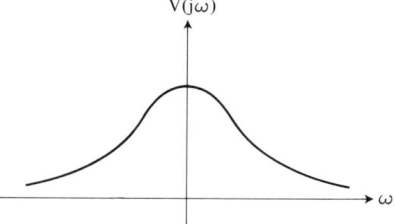

Fig. 3.3 A plot of $V(j\omega) = \mathscr{F}[f(t)e^{-\sigma t}]$.

For a pictorial view of the inverse $\mathscr{F}$- and $\mathscr{L}$-transforms, suppose that $V(j\omega)$ is as shown in Fig. 3.3. Another view of this same function is shown in Fig. 3.4(a). Then $F(\sigma_1 + j\omega)$ is pictured in Fig. 3.4(b). The shape of the two curves is identical, since $V(j\omega) = F(\sigma_1 + j\omega)$. The origin has simply been translated along the σ-axis by a distance σ_1.

To find the inverse $\mathscr{F}$-transform of $V(j\omega)$, we multiply by $e^{j\omega t}$ and integrate along the line $\sigma = 0$ from $\omega = -\infty$ to $\omega = +\infty$. This gives us $v(t)$, not $f(t)$.

To find the inverse $\mathscr{L}$-transform of $F(S)$, we multiply by e^{st} and integrate along the line $\sigma = \sigma_1$ from $\omega = -\infty$ to $\omega = +\infty$. This gives us $f(t)$.

Finally, let us compare the frequency functions.

1. The Fourier transform. The domain of $V(f)$ consists of real values of frequency. The range consists of complex values of V.

2. The Laplace transform. The domain of $V(S)$ consists of complex values of frequency. The range consists of complex values of V.

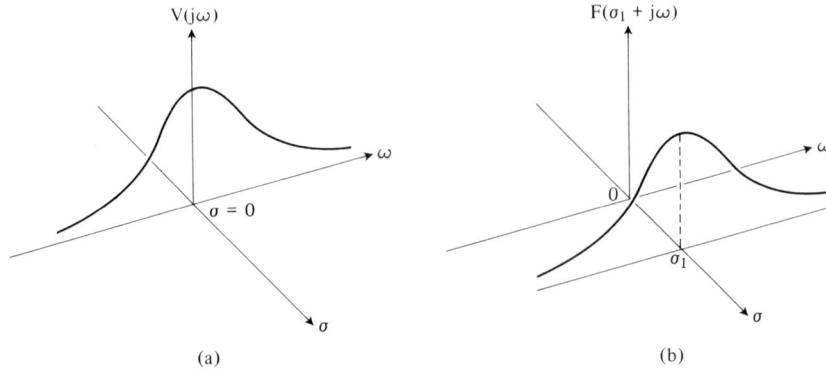

Fig. 3.4

3.3 PROPERTIES OF THE FOURIER TRANSFORM

The properties of the Fourier transform are identical to those of the Fourier series, with appropriate modifications for the continuous nature of the domain of $V(f)$. Again, the forward and inverse transforms are unique, and this is the meaning of the double-headed arrow.

$$\text{Uniqueness:} \quad v(t) \leftrightarrow V(f) \tag{3.11}$$

$$\text{Superposition:} \quad av(t) + bw(t) \leftrightarrow aV(f) + bW(f) \tag{3.12}$$

$$\text{Differentiation:} \quad \frac{d^n v(t)}{dt^n} \leftrightarrow (j\omega)^n V(f) \tag{3.13}$$

$$\text{Scaling:} \quad v(t/a) \leftrightarrow aV(af) \quad \text{for real positive } a \tag{3.14}$$

$$\text{Delay:} \quad v(t - t_0) \leftrightarrow V(f)e^{-j\omega t_0} \tag{3.15}$$

$$\text{Modulation:} \quad v(t)e^{j\omega_0 t} \leftrightarrow V(f - f_0) \tag{3.16}$$

$$\text{Convolution:} \quad \int_{-\infty}^{+\infty} v_1(\lambda)v_2(t-\lambda)\, d\lambda \leftrightarrow V_1(f)V_2(f) \tag{3.17}$$

$$\text{Multiplication:} \quad v_1(t)v_2(t) \leftrightarrow \int_{-\infty}^{+\infty} V_1(\xi)V_2(f-\xi)\, d\xi \tag{3.18}$$

The proofs of these properties follow closely those given in Chapter 2 for the corresponding properties of the series.

3.4 USE OF THE FOURIER TRANSFORM PROPERTIES

The differentiation property, Eq. (3.13), is useful in the evaluation of the transform of a pulse, just as this property was useful in the evaluation of the coefficients of the series. The procedure is identical to that illustrated in Section 2.6, so we will not repeat it here. Instead, we will illustrate the

3.4 USE OF THE FOURIER TRANSFORM PROPERTIES

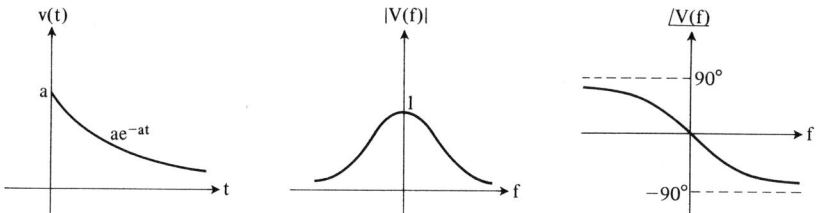

Fig. 3.5 The function ae^{-at} and its Fourier transform.

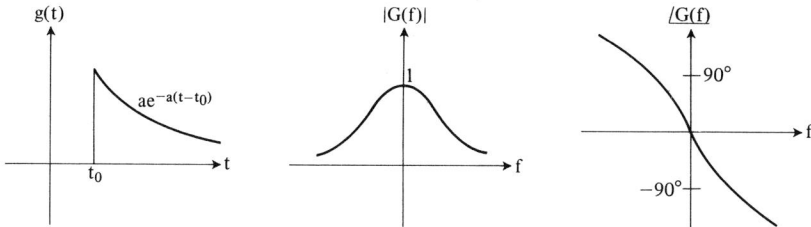

Fig. 3.6 The function $ae^{-a(t-t_0)}$ and its Fourier transform.

usefulness of the delay and modulation properties; these same techniques are applicable to the Fourier series.

Examples

1. *Use of the delay property.* Let $v(t)$ be a decaying exponential:

$$v(t) = \begin{cases} ae^{-at}, & t > 0, \\ 0, & t < 0. \end{cases} \quad (3.19)$$

Then the Fourier transform is

$$V(f) = \int_{-\infty}^{+\infty} v(t)e^{-j\omega t}\, dt = \int_{0}^{+\infty} ae^{-(a+j\omega)t}\, dt = \frac{a}{a+j\omega}.$$

Now suppose $v(t)$ is shifted along the time axis by t_0 seconds. Let $g(t) = v(t - t_0)$. Then $G(f)$ is related to $V(f)$ by $G(f) = V(f)e^{-j\omega t_0}$.

Notice the differences between Fig. 3.5 and Fig. 3.6. The magnitudes of $V(f)$ and $G(f)$ are identical. Shifting the function along the time axis changes only the phase of the corresponding frequency function.

2. *Use of the modulation property.* Let $v(t)$ be the decaying exponential of Example 1, Eq. (3.19). Now multiply $v(t)$ by a cosine waveform

$$y(t) = v(t) \cos \omega_0 t.$$

By Euler's formula, $y(t)$ is given by

$$y(t) = \frac{v(t)}{2} e^{j\omega_0 t} + \frac{v(t)}{2} e^{-j\omega_0 t}. \quad (3.20)$$

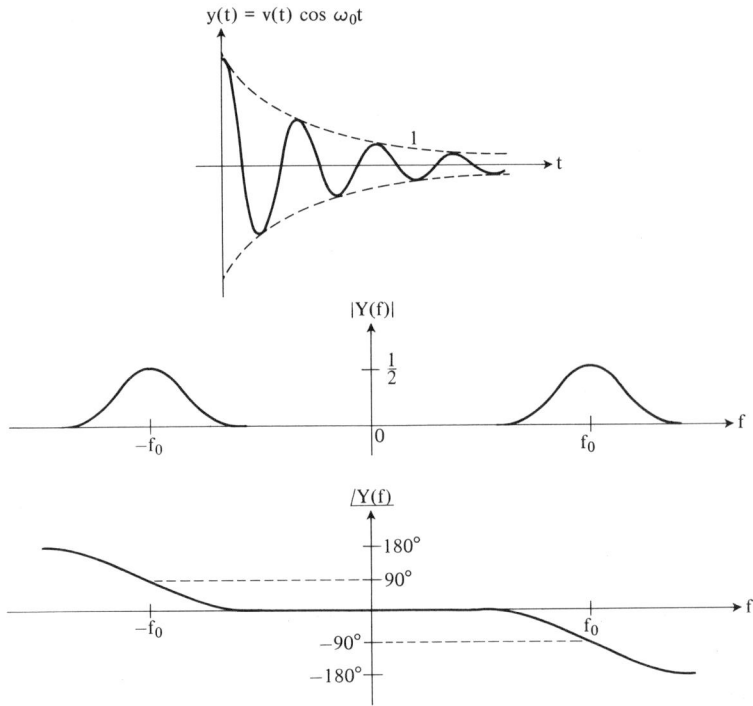

Fig. 3.7 The function $ae^{-at}\cos\omega_0 t$ and its Fourier transform.

Apply the modulation property to the first term of Eq. (3.20) to obtain

$$\frac{v(t)}{2}e^{j\omega_0 t} \leftrightarrow \tfrac{1}{2}V(f - f_0).$$

The second term is transformed into

$$\frac{v(t)}{2}e^{-j\omega_0 t} \leftrightarrow \tfrac{1}{2}V(f + f_0).$$

Now the superposition property yields

$$y(t) \leftrightarrow \tfrac{1}{2}V(f - f_0) + \tfrac{1}{2}V(f + f_0).$$

The magnitude and phase of $Y(f)$ are shown in Fig. 3.7, where $\omega_0 = 2\pi f_0$. If the frequency of the cosine wave were small (relative to the parameter a), then the magnitude spectrum $|Y(f)|$ would consist of a single broad hump. The phase spectrum would also be somewhat different. Figure 3.7 is the appropriate picture for $\omega_0 \gg a$.

The convolution and multiplication properties for both the series and transform are very useful, but illustrations of their usefulness require a knowledge of convolution, which is dealt with in the next chapter. We will have much more to say on this subject later.

3.5 BOUNDS ON THE SPECTRUM

In addition to its use in the evaluation of the series and transform of a signal, the differentiation property, Eq. (2.14) for the series and Eq. (3.13) for the transform, has another use. We may derive bounds on the magnitude of the frequency spectrum and relate these bounds to the time function. This might be useful in situations in which the task of finding the frequency spectrum is tedious and we need to know only the approximate spectrum. [We are using the term "frequency spectrum" to refer to the coefficients $V(f_k)$ it $v(t)$ is periodic and to the function $V(f)$ if $v(t)$ is aperiodic.]

First suppose $v(t)$ is periodic. The content, variation, and wiggliness of $v(t)$ are defined as

$$\text{content}_p = \frac{1}{T} \int_{t'}^{t'+T} |v(t)|\, dt, \tag{3.21}$$

$$\text{variation}_p = \frac{1}{T} \int_{t'}^{t'+T} \left|\frac{dv}{dt}\right| dt, \tag{3.22}$$

$$\text{wiggliness}_p = \frac{1}{T} \int_{t'}^{t'+T} \left|\frac{d^2v}{dt^2}\right| dt. \tag{3.23}$$

The frequency spectrum is related to the derivatives, according to Eq. (2.14), by

$$V_k = \frac{1}{T} \int_{t'}^{t'+T} v(t) e^{-j\omega_k t}\, dt, \tag{3.24}$$

$$j\omega_k V_k = \frac{1}{T} \int_{t'}^{t'+T} \frac{dv}{dt} e^{-j\omega_k t}\, dt, \tag{3.25}$$

$$-\omega_k^2 V_k = \frac{1}{T} \int_{t'}^{t'+T} \frac{d^2v}{dt^2} e^{-j\omega_k t}\, dt. \tag{3.26}$$

Now take absolute values of both sides of these equations and use the relationship

$$\left|\int_a^b f(t)g(t)\, dt\right| \leq \int_a^b |f(t)|\,|g(t)|\, dt \tag{3.27}$$

to obtain

$$|V_k| \leq \frac{1}{T} \int_{t'}^{t'+T} |v(t)|\, dt = \text{content}_p,$$

$$|\omega_k|\,|V_k| \leq \frac{1}{T} \int_{t'}^{t'+T} \left|\frac{dv}{dt}\right| dt = \text{variation}_p,$$

$$\omega_k^2\, |V_k| \leq \frac{1}{T} \int_{t'}^{t'+T} \left|\frac{d^2v}{dt^2}\right| dt = \text{wiggliness}_p,$$

48 THE FOURIER TRANSFORM **3.5**

or

$$|V_k| \leq \begin{cases} \text{content}_p, \\ \text{variation}_p/|\omega_k|, \\ \text{wiggliness}_p/\omega_k^2. \end{cases} \quad (3.28)$$

Next, suppose $v(t)$ is aperiodic (pulse). The content, variation, and wiggliness of $v(t)$ are defined as

$$\text{content}_a = \int_{-\infty}^{+\infty} |v(t)| \, dt, \quad (3.29)$$

$$\text{variation}_a = \int_{-\infty}^{+\infty} \left|\frac{dv}{dt}\right| dt, \quad (3.30)$$

$$\text{wiggliness}_a = \int_{-\infty}^{+\infty} \left|\frac{d^2v}{dt^2}\right| dt. \quad (3.31)$$

Note that the subscript p is used for periodic functions and the subscript a for aperiodic functions.

The relations between $V(f)$ and the derivatives of $v(t)$ are used in conjunction with relation (3.27) to obtain

$$|V(f)| \leq \begin{cases} \text{content}_a, \\ \text{variation}_a/|\omega|, \\ \text{wiggliness}_a/\omega^2. \end{cases} \quad (3.32)$$

The step-by-step procedure used to derive Eq. (3.32) is identical to that used to derive Eq. (3.28) from Eqs. (3.24), (3.25), and (3.26).

A less formal view of the content, variation, and wiggliness is as follows. The *content* is the total area (both positive and negative) under the curve of the function. The *variation* is the total change in altitude of the curve, that is, the total distance traveled by the curve over its range. The *wiggliness*, which is surely the most descriptive of these terms, is a measure of how much the curve wiggles, that is, the total change in the slope of the curve.

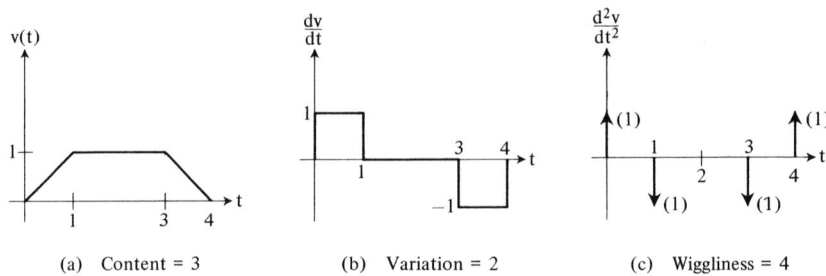

(a) Content = 3   (b) Variation = 2   (c) Wiggliness = 4

Fig. 3.8

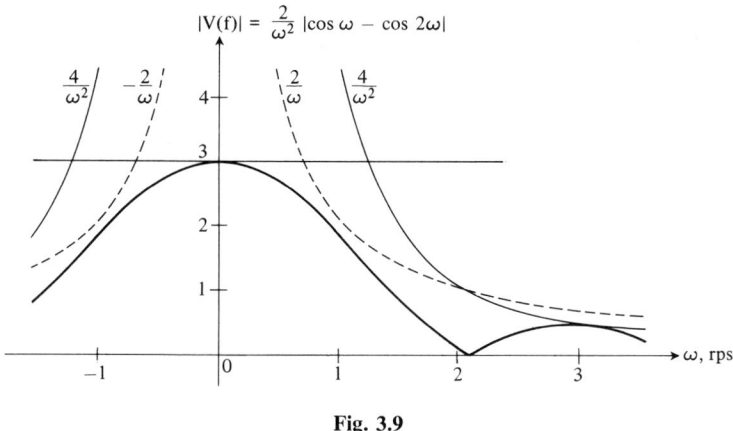

Fig. 3.9

Example 3. The function $v(t)$ and its derivatives are shown in Fig. 3.8. According to Eq. (3.32), the spectrum is bounded by

$$|V(f)| \leq \begin{cases} 3, \\ 2/|\omega|, \\ 4/\omega^2. \end{cases}$$

Figure 3.9 illustrates these bounds with the magnitude of $V(f)$ superimposed.

3.6 RAYLEIGH'S THEOREM AND ENERGY SPECTRAL DENSITY

Strictly speaking, the term "Parseval's theorem" refers to Eq. (2.24). The corresponding relation for energy signals was first used by Lord Rayleigh in his study of blackbody radiation; hence the above title.

Rayleigh's theorem relates the energy in the frequency domain to the energy in the time domain, just as Parseval's theorem is a relation involving power. Recall that the equation for energy is

$$E = \int_{-\infty}^{+\infty} |v(t)|^2 \, dt = \int_{-\infty}^{+\infty} v^*(t)v(t) \, dt. \tag{1.2}$$

Now replace $v(t)$ by its Fourier transform, Eq. (3.2), to obtain

$$E = \int_{-\infty}^{+\infty} v^*(t) \left[\int_{-\infty}^{+\infty} V(f) e^{j2\pi ft} \, df \right] dt.$$

Again (as in Section 2.7) we change the order of integration and get

$$E = \int_{-\infty}^{+\infty} V(f) \left[\int_{-\infty}^{+\infty} v^*(t) e^{j2\pi ft} \, dt \right] df.$$

The term in brackets is $V^*(f)$, the complex conjugate of $V(f)$. Thus the energy is

$$E = \int_{-\infty}^{+\infty} |v(t)|^2 \, dt = \int_{-\infty}^{+\infty} |V(f)|^2 \, df, \tag{3.33}$$

which is known as Rayleigh's theorem.

Energy Spectral Density

Just as the power is distributed along the frequency axis (see Section 2.7), so must the energy be distributed along the frequency axis. That is, when dealing with an energy signal, we should be able to rig up some sort of apparatus to measure this energy in narrow frequency bands, and thus obtain a plot of energy versus frequency.

The energy spectral density $W(f)$ is the function (1) that describes the relative amount of energy versus frequency, and (2) whose total area under $W(f)$ is the total energy. The symbol $W(f)$ is used to avoid confusion with E, the total energy. Thus we have

$$E = \int_{-\infty}^{+\infty} W(f) \, df.$$

The energy in the frequency band $f_1 < f < f_2$ is given by

$$E_{f_1 < f < f_2} = \int_{f_1}^{f_2} W(f) \, df.$$

We are interested in the relationship between $W(f)$ and the quantity $|V(f)|^2$ of Eq. (3.33). It should be obvious after the following discussion that

$$W(f) = |V(f)|^2, \tag{3.34}$$

where the total area under $W(f)$ is the energy in $v(t)$.

We defined the energy spectral density $W(f)$ by two properties:

1. $W(f)$ describes the relative amount of energy versus frequency.
2. The total area under $W(f)$ is the energy.

The function $|V(f)|^2$ satisfies property 1, and Rayleigh's theorem, Eq. (3.33), states that the total area under $|V(f)|^2$ is the energy. We conclude that the function $|V(f)|^2$ fills the bill for the energy spectral density function. Hence Eq. (3.34) is valid.

Example 4. The square pulse shown in Fig. 3.10(a) has energy

$$E = \int_{-\infty}^{+\infty} |v(t)|^2 \, dt = \int_0^1 1 \, dt = 1.$$

The Fourier transform of $v(t)$ is

$$V(f) = \int_{-\infty}^{+\infty} v(t) e^{-j\omega t} \, dt = e^{-j\omega/2} \left[\frac{\sin \omega/2}{\omega/2} \right].$$

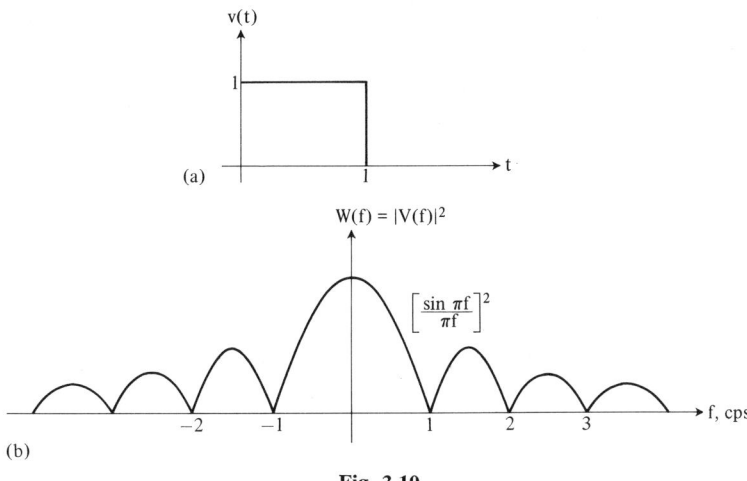

Fig. 3.10

The total energy is distributed in frequency according to

$$|V(f)|^2 = \left[\frac{\sin \omega/2}{\omega/2}\right]^2,$$

which is shown in Fig. 3.10(b).

3.7 EXAMPLES

We now provide some examples of the use of the preceding theory in applications to circuits, as we did in Section 2.8 for the theory of Fourier series.

5. Find the Fourier transform of $f_1(t)$, shown in Fig. 3.11. The function $f_1(t)$ is given by

$$f_1(t) = \begin{cases} e^{at}, & t < 0, \\ e^{-at}, & t > 0. \end{cases}$$

Then

$$F_1(f) = \int_{-\infty}^{0} e^{at}e^{-j\omega t}\,dt + \int_{0}^{\infty} e^{-at}e^{-j\omega t}\,dt = \frac{1}{a - j\omega} + \frac{1}{a + j\omega} = \frac{2a}{a^2 + \omega^2}.$$

(See Fig. 3.12.)

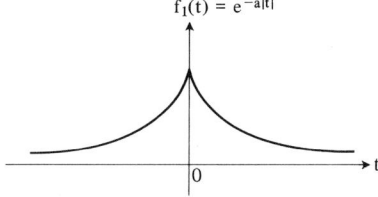

Fig. 3.11

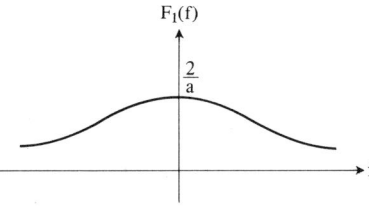

Fig. 3.12

Note. The frequency function $F_1(f)$ is real. This is true because $f_1(t)$ is even. Generally, the Fourier transform of a time function is a complex function of frequency. Therefore, to graph the function, we will need either a three-dimensional plot or two separate plots.

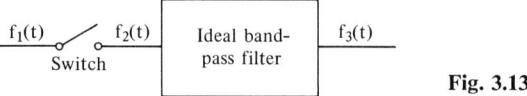

Fig. 3.13

6. The waveform in Example 5 with $a = 2$ is passed through the circuit shown in Fig. 3.13. The switch opens and closes at a rate of 10 cps. That is, the switch is periodic with period $T = \frac{1}{10}$ sec. The bandpass filter has unity gain and zero phase shift with a bandwidth of 6 cps centered at 10 cps.

a) Sketch the time function $f_2(t)$ and its spectrum.

b) Sketch the time function $f_3(t)$.

The function $f_2(t)$ can be thought of as the product of $s(t)$ (shown in Fig. 3.14) and $f_1(t)$. Therefore, by the multiplication property, Eq. (3.18), the spectrum of $f_2(t)$ is given by the convolution of $F_1(f)$ with $S(f)$. That is,

$$f_2(t) = f_1(t) \cdot s(t) \leftrightarrow F_1(f) * S(f) = F_2(f).$$

The spectrum of $s(t)$ has already been found in Example 4 of Chapter 2. This spectrum is shown in Fig. 3.15(a). Figure 3.15(b) shows the spectrum of

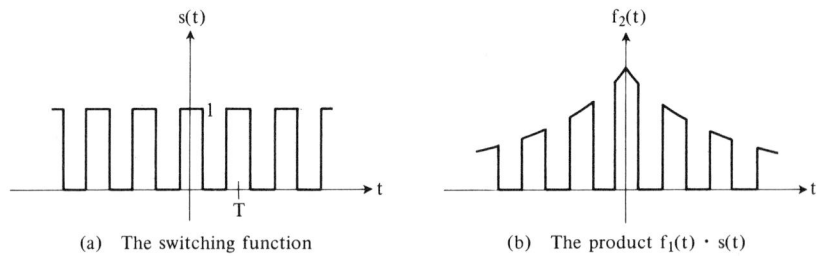

(a) The switching function (b) The product $f_1(t) \cdot s(t)$

Fig. 3.14

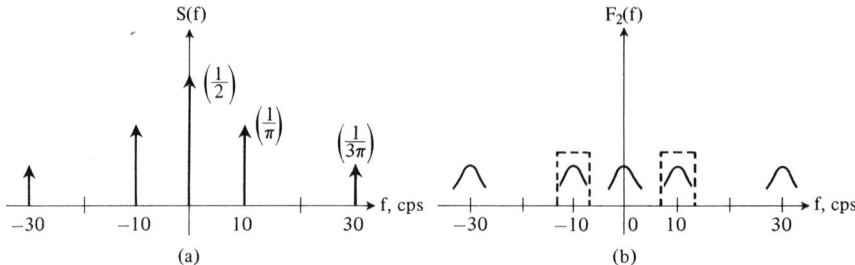

Fig. 3.15

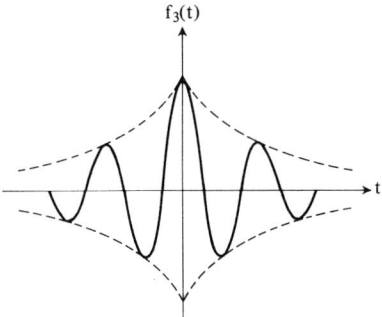

Fig. 3.16

$f_2(t)$. (This result is discussed in the note following the example.) The dashed line in Fig. 3.15(b) is the filter characteristic. The filter output $F_3(f)$ is the two humps inside the dashed lines. This output is recognized (see method 1 below) as the product in the time domain of $f_1(t)$ and a sinusoid. (See Fig. 3.16.)

The time function $f_3(t)$ is given by

$$f_3(t) = f_1(t) \cdot \frac{2}{\pi} \cos 2\pi(10)t. \tag{3.35}$$

There are two slightly different ways to derive Eq. (3.35).

Method 1. The switching function $s(t)$ can be written in trigonometric form by use of Euler's formula. Thus we get

$$s(t) = \frac{1}{2} + \frac{2}{\pi} \cos 2\pi(10)t + \frac{2}{3\pi} \cos 2\pi(30)t + \cdots$$

Therefore $f_2(t)$, which is the product $f_1(t)s(t)$, is

$$f_2(t) = \tfrac{1}{2}f_1(t) + f_1(t) \cdot \frac{2}{\pi} \cos 2\pi(10)t + f_1(t) \cdot \frac{2}{3\pi} \cos 2\pi(30)t + \cdots \tag{3.36}$$

The filter output is composed of frequencies centered around 10 cps, and this is the second term on the right-hand side of Eq. (3.36).

Method 2. Consider the product of the two time functions in Eq. (3.35). From the multiplication property, Eq. (3.18), the spectrum $F_3(f)$ is the convolution of the spectrum $F_1(f)$ with the spectrum for $(2/\pi) \cos 2\pi(10)t$. This is just the output of the filter.

Notes. a) Equation (3.35) is only an approximation. The output of the filter will contain some energy from the components in Eq. (3.36) at zero frequency and at 30 cps. Also, not all of the energy in the 10-cps term is within the filter bandwidth.

b) The spectrum $F_2(f)$ in Fig. 3.15(b) is derived by convolving each of the δ-functions in Fig. 3.15(a) with the spectrum $F_1(f)$. Since convolution is

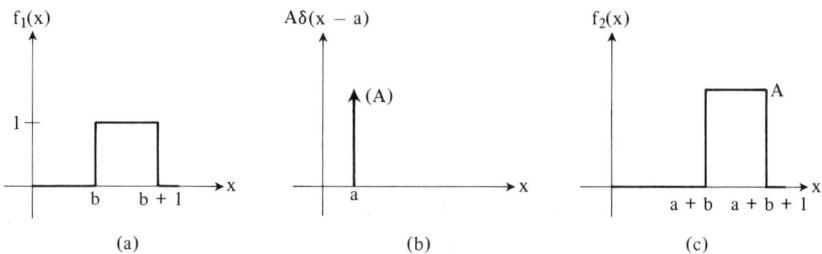

Fig. 3.17

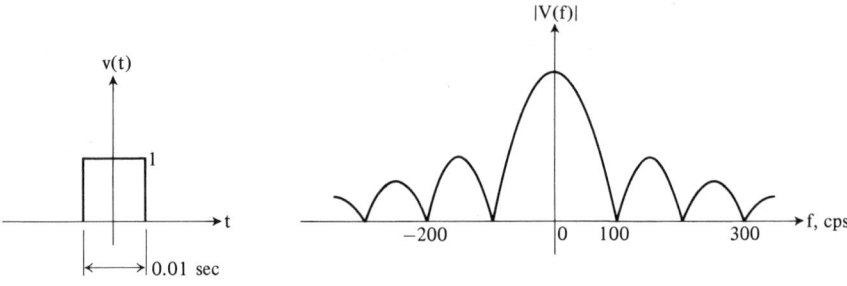

Fig. 3.18

discussed in the next chapter, the reader who is unfamiliar with this subject must take this result on faith. Convolution with a δ-function is a simple operation, however, and the following rule of thumb is presented for justification of Fig. 3.15(b).

Suppose $f_1(x)$ is to be convolved with $A\delta(x - a)$, shown in Fig. 3.17(a) and (b). Then the result of this convolution is just $f_1(x)$ shifted a units along the x-axis and multiplied by the number A (Fig. 3.17c).

7. The energy signal $v(t)$ and its magnitude spectrum are shown in Fig. 3.18. This signal is applied to an ideal low-pass filter with a bandwidth of 10 kcs and unity gain.

a) Compute the energy in the input signal on a one-ohm basis.
b) Estimate the energy of the output signal on a one-ohm basis.

The energy in the input signal is

$$E = \int_{-\infty}^{+\infty} |v(t)|^2 \, dt = 0.01 \text{ V}^2\text{-sec.}$$

Since the filter bandwidth is 100 times larger than the first zero of $|V(f)|$, there is very little energy that does not get through the filter. Therefore the energy in the output signal is approximately 0.01 V²-sec.

8. Figure 3.19 shows several time functions and their corresponding spectra. Use these illustrations as practice problems.

3.7 EXAMPLES 55

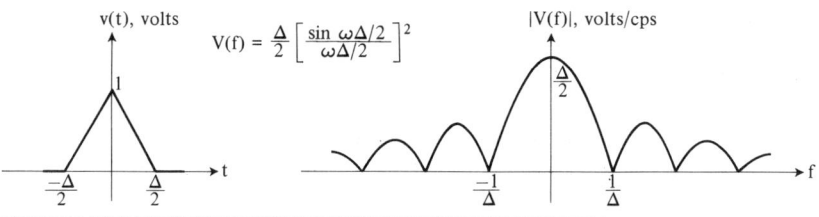

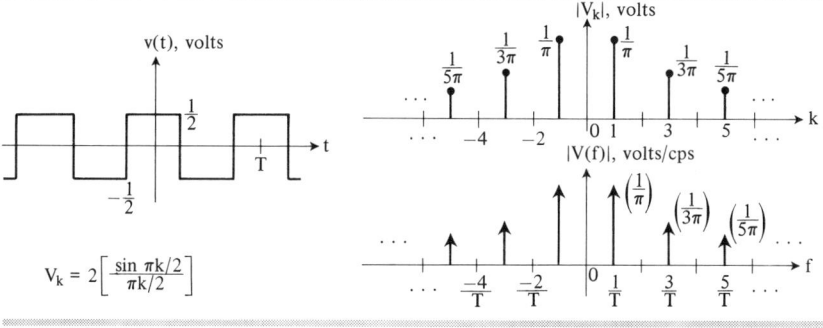

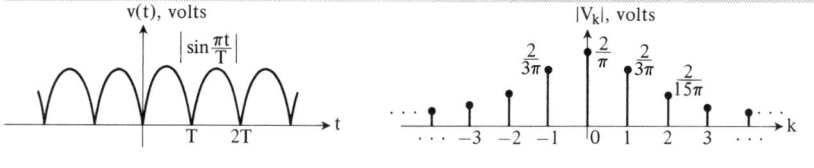

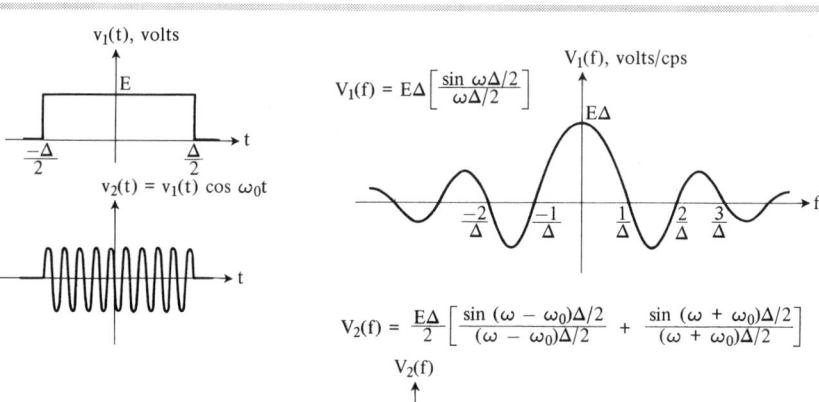

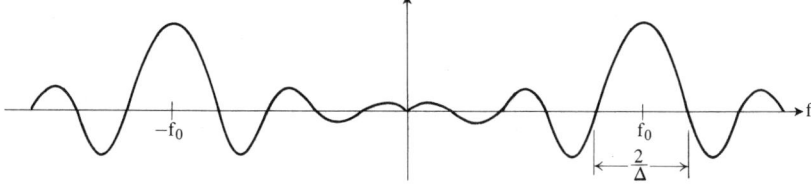

Fig. 3.19

PROBLEMS

1. Find and compare the Fourier transforms of each of the functions shown in Fig. 3.20.
2. a) Find $V_1(f)$, the spectrum for $v_1(t)$ in Fig. 3.21(a).
   b) Using the results of part (a), find $V_2(f)$ for the waveform in Fig. 3.21(b).

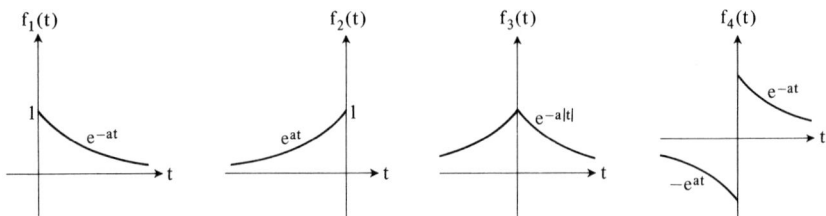

Fig. 3.20

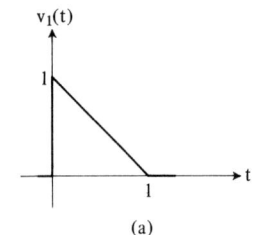
(a)

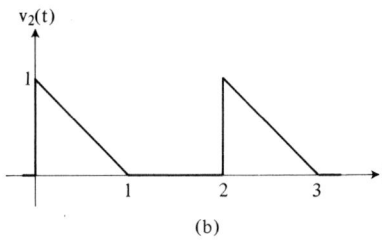
(b)

Fig. 3.21

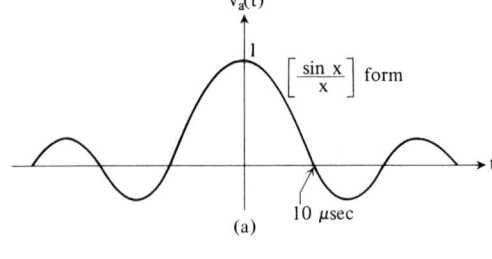
(a)

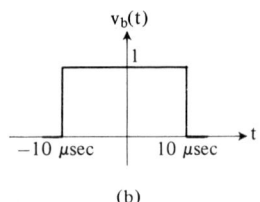
(b)

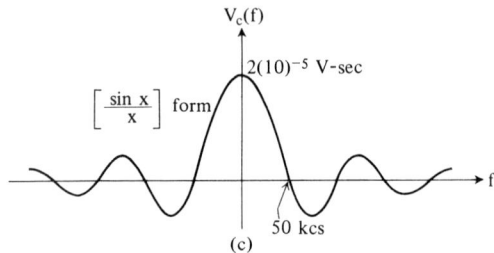
(c)

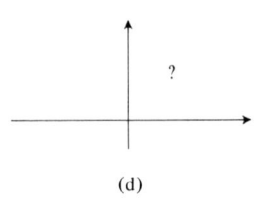
(d)

Fig. 3.22

3. Prove the properties of the Fourier transform given in Section 3.3 (except uniqueness).
4. a) Find the spectrum of the time function shown in Fig. 3.22(a).
b) Find the spectrum of the time function shown in Fig. 3.22(b).
c) Find the time function whose spectrum is shown in Fig. 3.22(c).
d) For Fig. 3.22(d), what is the question and what is the answer?
5. A signal $s(t)$ has Fourier transform $S(f) = |S(f)| e^{j\underline{/S(f)}}$, with $|S(f)|$ and $\underline{/S(f)}$ as shown in Fig. 3.23. Find and sketch $s(t)$.

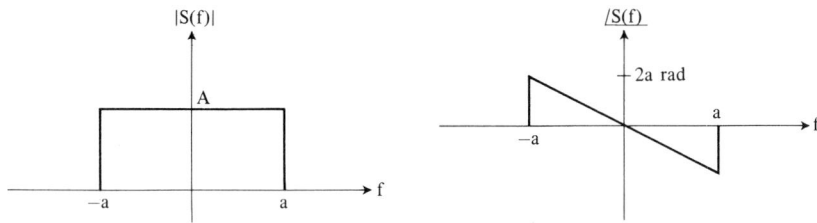

Fig. 3.23

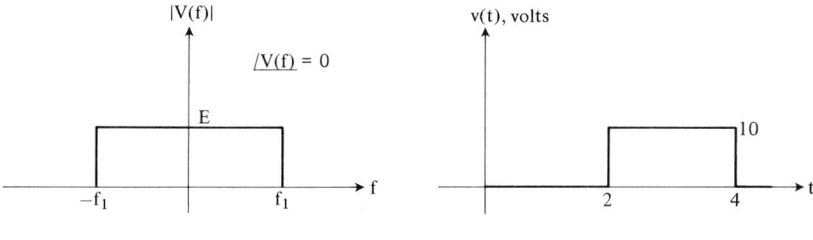

Fig. 3.24   Fig. 3.25

6. The $V(f)$ shown in Fig. 3.24 is the Fourier transform of $v(t)$.
a) Find the Fourier transform of $g(t) = [v(t)]^2$.
b) Find the Fourier transform of $h(t) = v(t - 2)$.
7. The Fourier transform of $f(t)$ is $F(f)$, and the Fourier transform of $g(t)$ is $G(f)$. For each of the following cases find $G(f)$ in terms of $F(f)$.
a) $g(t) = af(t)$
b) $g(t) = f(bt)$
c) $g(t) = f(t + c)$
8. The transform of a square pulse is given in the practice problems of Example 8, Section 3.7. Use this and the results of Problem 7 to find the Fourier transform of the pulse shown in Fig. 3.25.
9. a) Find the Fourier transform of the waveform $v_1(t)$ shown in Fig. 3.26.
b) The waveform $v_1(t)$ is multiplied by $\cos \omega_0 t$. Find and sketch the spectrum.
c) Repeat (b) with $v_1(t)$ multiplied by $\sin \omega_0 t$, $\omega_0 \gg 1/\Delta$.

58   THE FOURIER TRANSFORM

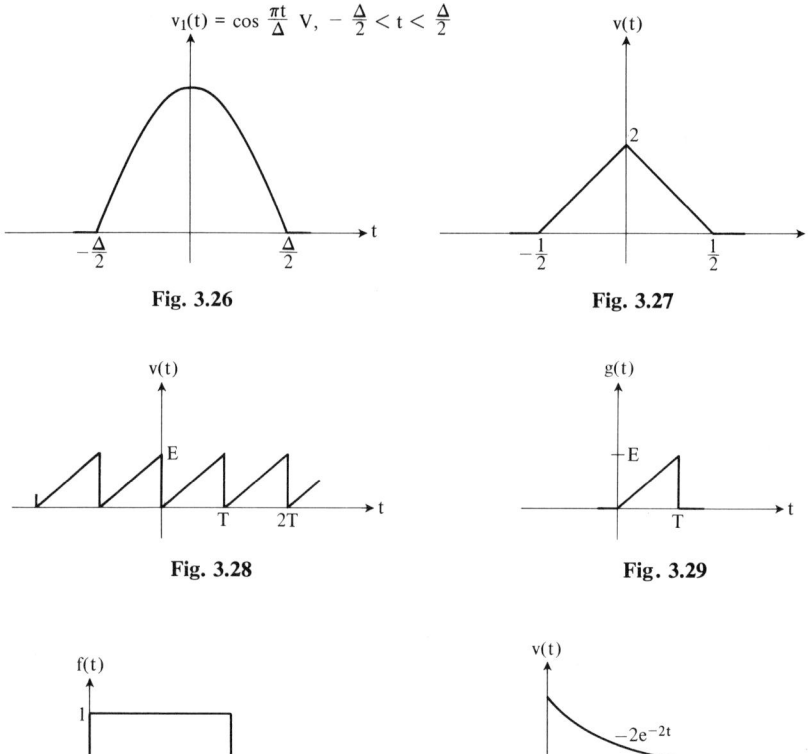

Fig. 3.26

Fig. 3.27

Fig. 3.28

Fig. 3.29

Fig. 3.30

Fig. 3.31

10. For the waveform shown in Fig. 3.27:
  a) find $V(f)$;
  b) sketch, with dimensions, $|V(f)|$;
  c) determine the content, variation, and wiggliness for $v(t)$;
  d) sketch the content, variation/$|\omega|$, and wiggliness/ω^2 on the answer to part (b).
11. The integral I is given by
$$I = \int_{-\infty}^{+\infty} \left| \frac{d^3v(t)}{dt^3} \right| dt,$$
where $v(t)$ is a pulse signal. How is I related to $V(f)$, the Fourier transform of $v(t)$? What can I be used for?
12. Find the bounds on the magnitude spectrum for the periodic sawtooth waveform shown in Fig. 3.28. Compare your answer to the results given in Example 11 of Chapter 2.

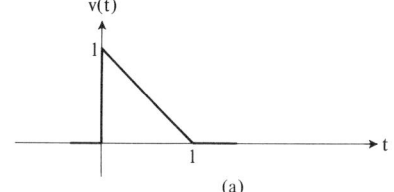
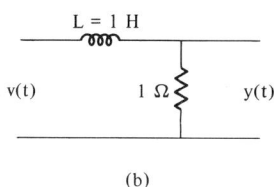

Fig. 3.32

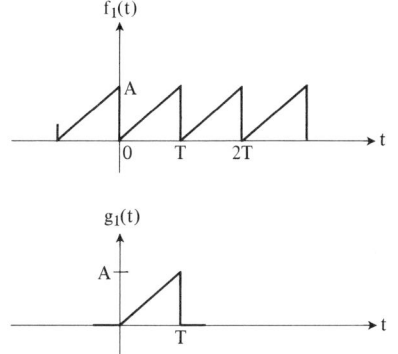
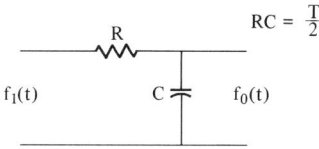
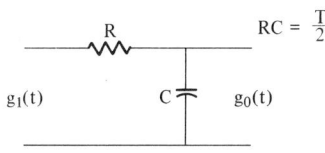

Fig. 3.33

13. Find and plot bounds on the magnitude spectrum for the waveform $g(t)$ shown in Fig. 3.29. Compare your answer to the results of Problem 12.
14. Find and plot the energy spectral density for the waveform shown in Fig. 3.29.
15. Find and plot the energy spectral density for the waveforms shown in Fig. 3.21.
16. The pulse signal $f(t)$ shown in Fig. 3.30 is supplied to an ideal low-pass filter with cutoff frequency f_m. Find the approximate energy in the output of the filter for the following values of f_m.
a) $f_m = 0.1$ cps
b) $f_m = 1$ cps
c) $f_m = 10$ cps
17. Find and plot the energy spectral density for the signal shown in Fig. 3.31.
18. The voltage $v(t)$ shown in Fig. 3.32(a) is applied to the low-pass filter shown in Fig. 3.32(b). Find the approximate energy in the output $y(t)$ on a one-ohm basis.
19. Find and plot the energy spectral density for each of the waveforms shown in Fig. 3.20.
20. Determine the amplitude and phase spectra of $f_0(t)$ and $g_0(t)$ shown in Fig. 3.33.

FURTHER READING

1. W. KAPLAN, *Operational Methods for Linear Systems*, Addison-Wesley, Reading, Mass., 1962.
2. A. PAPOULIS, *The Fourier Integral*, McGraw-Hill, New York, 1962.
3. B. P. LATHI, *Signals, Systems, and Communications*, Wiley, New York, 1965.
4. S. J. MASON and H. J. ZIMMERMAN, *Electronic Circuits, Signals, and Systems*, Wiley, New York, 1960.

All these texts provide a thorough discussion of the material in this chapter. The material in Section 3.5 (bounds on the spectrum) is due to Mason and Zimmerman, Section 6.18.

4
SYSTEMS

4.1 CLASSIFICATION OF SYSTEMS

Conventionally, a linear system is described by a linear differential equation, the system function $H(s)$ in the frequency domain, or the impulse response $h(t)$ in the time domain. In this chapter we briefly review these representations and their relationship to each other. We assume that the system is not time-variant.

There is an equivalence between systems and functions. A system has some input $r_1(t)$ and some output $r_2(t)$ (at least the systems we are interested in have an input and output). Our function machine of Chapter 2 also has an input $x(t)$ and the corresponding output $y(t)$. This equivalence is shown in Fig. 4.1. If $x(t)$ and $y(t)$ are both functions, then the function machine must be an operator.

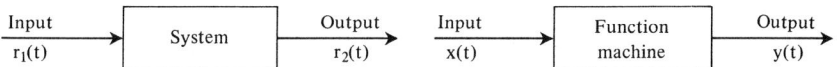

Fig. 4.1 The equivalence of physical systems and operators.

In system analysis, one of the major problems is to arrive at a mathematical model for a given system. By a mathematical model we mean a mathematical expression of the relationship between input and output (the function machine) that approximates the physical system. That is, if $x(t)$ approximates the system input $r_1(t)$, then $y(t)$ should approximate $r_2(t)$, regardless of whether our model is given in the form of a differential equation, a system function, or even a simple table relating input and output.

Students are often not aware of this problem, since traditionally they are given the mathematical model of a system rather than the physical system. It is safe to say that many students are not even aware of a difference between mathematical models and physical systems (through no fault of their own).

Systems can be classified into several different categories. We will discuss only the four following ways of classifying systems:
1. deterministic or nondeterministic,
2. time-invariant or time-variant,
3. linear or nonlinear,
4. with memory or without memory.

We shall now discuss the meaning of these terms. Recall that the mathematical model for a system is an operator (Fig. 4.1). That is, the input (an element of the domain) is a function, and the output (an element of the range) is a function. Let us use the letter L to denote the operator. Then the system output y is given in terms of the system input x by

$$y = L[x]. \tag{4.1}$$

Notes. a) If we are in the frequency domain, then the input is $X(s)$, the system transfer function is $H(s)$, and the output $Y(s)$ is given by

$$Y(s) = H(s) X(s). \tag{4.2}$$

In this case the operator L corresponds to multiplication by $H(s)$.

b) If the input $x(t)$, the output $y(t)$, and the system function $h(t)$ are given as functions of time, then the output $y(t)$ is given by

$$y(t) = \int_{-\infty}^{+\infty} h(t - \lambda) x(\lambda) \, d\lambda, \tag{4.3}$$

which is the convolution operation (to be discussed in Section 4.4). Now the operator L corresponds to multiplying the input by a function h and integrating the product, as in Eq. (4.3).

c) The differential equation that describes the system might be something like

$$x(t) = a_n \frac{d^n y}{dt^n} + a_{n-1} \frac{d^{n-1} y}{dt^{n-1}} + \cdots + a_1 \frac{dy}{dt} + a_0 y. \tag{4.4}$$

This equation describes the relationship between the input $x(t)$ and the output $y(t)$, and hence is an operator.

1. *Deterministic or nondeterministic.* A system is random (nondeterministic) if the differential equation that characterizes the system has any random coefficients. Equivalently, a nondeterministic system has one or more random parameters. Thus the system characterized by Eq. (4.4) is random if any one of the coefficients $a_n, \ldots, a_1, a_0$ is random. Otherwise the system is deterministic. [Here $x(t)$ is the input and $y(t)$ is the output.] In this text we will deal with only deterministic systems.

2. Time-invariant or time-variant.
A system with an input-output relationship given by Eq. (4.1) is time-invariant if

$$y(t + \epsilon) = L[x(t + \epsilon)] \quad \text{for all } \epsilon. \tag{4.5}$$

This implies that the response to the input is independent of the time at which the input is applied. An example of a time-variant system is the simple circuit shown in Fig. 4.2. The capacitor is variable, and if it is driven by a motor, then the response $y(t)$ will depend on the manner in which C changes with time. In this text we will deal with only time-invariant systems.

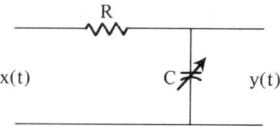

Fig. 4.2

3. Linear or nonlinear.
A system with an input-output relationship given by Eq. (4.1) is linear if

$$L[a_1 x_1(t) + a_2 x_2(t)] = a_1 L[x_1(t)] + a_2 L[x_2(t)]. \tag{4.6}$$

This definition really embodies two different criteria. If $a_1 = a_2 = 1$, then

$$L[x_1(t) + x_2(t)] = L[x_1(t)] + L[x_2(t)]. \tag{4.7}$$

This property is called *additivity*. If $x_2(t) = 0$, then

$$L[a_1 x_1(t)] = a_1 L[x_1(t)]. \tag{4.8}$$

This property is called *homogeneity*. Therefore a system is linear if it is both additive and homogeneous. Otherwise it is nonlinear.

A system with an input-output relationship given by

$$y(t) = L[x(t)] = x^2(t) \tag{4.9}$$

is nonlinear. The system is neither additive nor homogeneous.

Many systems are linear for some range of their operation, but nonlinear for some values of a_1, a_2, $x_1(t)$, and $x_2(t)$. A vacuum-tube amplifier in class-A operation is approximately linear if the magnitude of the input is restricted, but is driven into saturation or cutoff if the input voltage becomes too large.

4. With memory or without memory.
A system with an input-output relationship given by Eq. (4.1) is without memory if

$$y(t) = g[x(t)], \tag{4.10}$$

where g is a function of only x. Therefore $y(t_1) = g[x(t_1)]$ and $y(t_1)$ does not depend on $x(t)$ at $t \neq t_1$. Generally, a system with energy-storage elements has memory (though special configurations of the elements can cause us to make an exception to this rule). In electric circuits a system composed of resistors (and possibly amplifiers) has no memory. That is, a simple gain circuit has no memory.

4.2 TYPES OF RESPONSE

Consider a general system S with input $x(t)$ and output $y(t)$, as indicated in Fig. 4.3(a). In more familiar terms, this might be a circuit in which the input is the applied voltage and the output is the resulting current. For the RLC-circuit shown in Fig. 4.3(b), the relationship between the input voltage $v(t)$ and the output current $i(t)$ is given by

$$\frac{dv}{dt} = L\frac{d^2 i}{dt^2} + R\frac{di}{dt} + \frac{1}{C} i(t). \tag{4.11}$$

Therefore, for the general system S the relationship between the input $x(t)$ and the output $y(t)$ should be something like

$$b_0 x(t) + b_1 \frac{dx}{dt} + \cdots + b_k \frac{d^k x}{dt^k} = a_0 y(t) + a_1 \frac{dy}{dt} + \cdots + a_n \frac{d^n y}{dt^n}, \tag{4.12}$$

where S is an nth-order system.

The total solution $y(t)$ for this differential equation is the sum $y_{sf}(t) + y_f(t)$, where $y_{sf}(t)$ is the source-free (complementary) solution found by setting $x(t) = 0$, and $y_f(t)$ is the forced response (particular solution) related to the forcing function. To find the source-free solution to Eq. (4.12), we set the forcing function equal to zero, and obtain

$$0 = a_0 y(t) + a_1 \frac{dy}{dt} + \cdots + a_n \frac{d^n y}{dt^n}. \tag{4.13}$$

Substituting the trial solution $y_{sf}(t) = A e^{st}$ into Eq. (4.13), we get

$$0 = a_0 + a_1 s + \cdots + a_n s^n. \tag{4.14}$$

This is called the *characteristic equation*. There are n values of s that satisfy the characteristic equation (some of which may be repeated), and these n values of s are, in general, complex numbers.

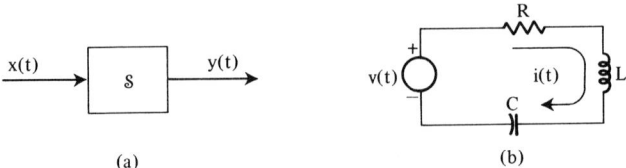

Fig. 4.3

4.2 TYPES OF RESPONSE

The source-free solution consists of n terms corresponding to the n roots $s_1, s_2, \ldots, s_n$ of the characteristic equation. This solution is of the form

$$y_{sf}(t) = A_1 e^{s_1 t} + A_2 e^{s_2 t} + \cdots + A_n e^{s_n t}, \qquad (4.15)$$

provided there are no repeated roots. If Eq. (4.14) does have repeated roots, the solution is of slightly different form, but there are still n terms.

The point here is that the roots of the characteristic equation are directly related to the source-free solution.

Note. We have established the following:

1. $y(t) = y_{sf}(t) + y_f(t)$. \hfill (4.16)
2. The roots of the characteristic equation are the exponents in the source-free solution.

There is an alternative method of separating the response $y(t)$ into two parts, namely, the transient response and the steady-state response. Then

$$y(t) = y_t(t) + y_{ss}(t), \qquad (4.17)$$

where $y_t(t)$ is the transient response and $y_{ss}(t)$ is the steady-state response.

The transient response is that portion of $y(t)$ that decays with time (or else increases without bound). The steady-state response is either constant or periodic with time.

Note. The transient response is often identical to the source-free response [therefore, $y_{ss}(t) = y_f(t)$]. For example, in Fig. 4.3(b) if the applied voltage is a constant (battery), then the forced response is zero, and this is the steady-state response. The transient response is then the source-free response. But this is not always the case.

Example 1. Consider a linear electric circuit S. The roots of the characteristic equation for S are shown in Fig. 4.4. For a given input applied at $t = 0$,

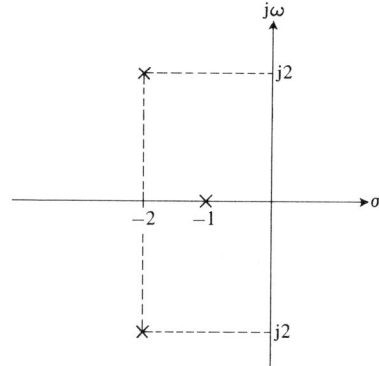

Fig. 4.4

the output of S is

$$y(t) = \tfrac{1}{2}e^{-t} + 3\sin 2t + 3 + 4e^{-3t}\cos 2t. \quad (4.18)$$

Find

a) the steady-state response,
b) the transient response,
c) the source-free response,
d) the forced response.

Solution

a) $y_{ss}(t) = 3 + 3\sin 2t$
b) $y_t(t) = \tfrac{1}{2}e^{-t} + 4e^{-3t}\cos 2t$
c) $y_{sf}(t) = \tfrac{1}{2}e^{-t}$
d) $y_f(t) = 3\sin 2t + 3 + 4e^{-3t}\cos 2t$

4.3 THE SYSTEM FUNCTION $H(s)$ IN THE FREQUENCY DOMAIN

Consider a time-invariant system. Sometimes the system function $H(s)$ is defined simply as the ratio of output to input:

$$H(s) = \frac{Y(s)}{X(s)}. \quad (4.19)$$

At other times $H(s)$ is defined as the response to an exponential time function divided by the exponential time function:

$$H(s) = \frac{\text{response to } e^{st}}{e^{st}}. \quad (4.20)$$

The two definitions are equivalent for certain values of s. Let us now investigate Eqs. (4.19) and (4.20) to determine (1) for which values of s the two are equivalent, and (2) what $H(s)$ can be used for.

The Signal e^{st}

In general, s is a complex number:

$$s = \sigma + j\omega. \quad (4.21)$$

Therefore the signal e^{st} may be put in the form

$$e^{st} = e^{\sigma t}e^{j\omega t} = e^{\sigma t}(\cos \omega t + j \sin \omega t). \quad (4.22)$$

This signal cannot actually be generated in the laboratory, but consider the first term, $e^{\sigma t} \cos \omega t$. Figure 4.5 illustrates the various waveforms obtained for different values of s. If s is a real number (Fig. 4.5a), the corresponding $f(t)$ is an exponential $e^{\sigma t}$. If s is complex, $f(t)$ is sinusoidal, with envelope equal to $e^{\sigma t}$, as in Fig. 4.5(b), (c), and (d).

4.3 THE SYSTEM FUNCTION $H(s)$ IN THE FREQUENCY DOMAIN

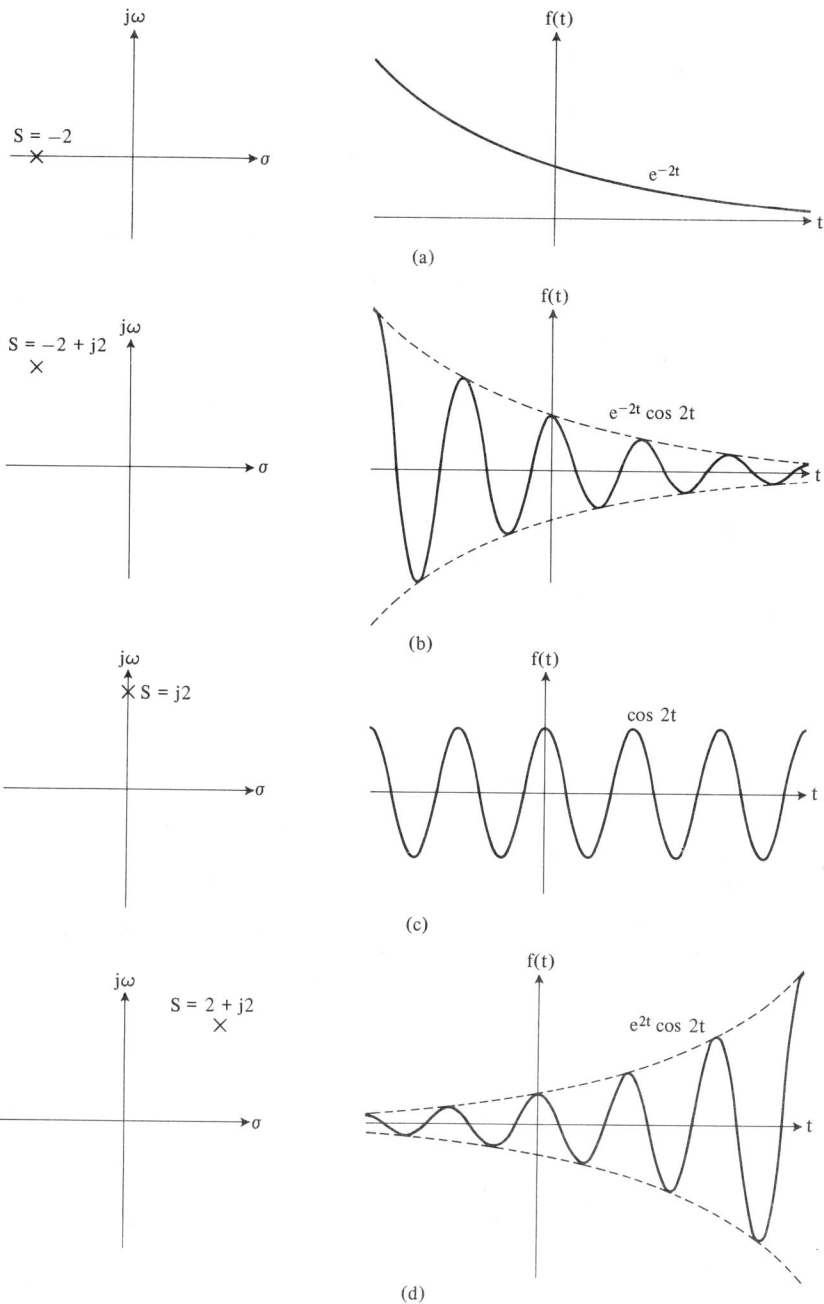

Fig. 4.5

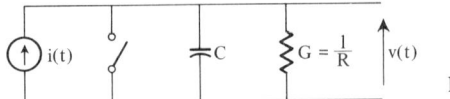

Fig. 4.6

Now apply the signal e^{st} to a system. Consider, for example, the system shown in Fig. 4.6. The input is current, $i(t) = e^{st}$, and the output is the voltage $v(t)$ across the conductor G. Therefore $H(s)$ is the ratio $V(s)/I(s)$. (Note that it is not necessary that $H(s)$ be the ratio of two voltages, though we are most often interested in the ratio of output voltage to input voltage.)

If the switch is opened at $t = 0$, then the current applied to the circuit is

$$i(t) = \begin{cases} 0, & t < 0, \\ e^{st}, & t \geq 0. \end{cases} \qquad (4.23)$$

Kirchhoff's current law can be used to obtain the differential equation

$$i(t) = Gv(t) + C\frac{dv}{dt}. \qquad (4.24)$$

Solving for $v(t)$, we obtain

$$v(t) = Ae^{-(G/C)t} + \left(\frac{1}{G + sC}\right)e^{st}, \qquad t \geq 0, \qquad (4.25)$$

which is the sum of two terms. The first term is the source-free response (complementary solution) and is characteristic of the circuit alone. The second term is the forced response (particular solution) and is dependent on the circuit and the source. The constant A is found from initial conditions in the system.

The two terms in Eq. (4.25) are exponentials. If

$$\sigma = \mathrm{Re}\,(s) > -G/C, \qquad (4.26)$$

then the source-free term will decay faster than the forced term. Otherwise, the forced response will decay faster. Equation (4.26) is known as the *dominance condition*. After a very long time, the voltage $v(t)$ will be given by the particular solution, and thus the e^{st}-term dominates the $e^{-(G/C)t}$-term.

Now suppose the exponential e^{st} is applied to the system at $t = -\infty$. Then at any finite time the output $v(t)$ is given by the second term of Eq. (4.25) so long as the dominance condition, Eq. (4.26), holds. This is the *forced response*, and according to Eq. (4.20), $H(s)$ is given by

$$H(s) = \frac{1}{G + sC}, \qquad \sigma > -\frac{G}{C}. \qquad (4.27)$$

For values of σ less than $-G/C$, Eqs. (4.20) and (4.19) yield different results. Therefore, if the real part of s is large enough so that the source-free response decays faster than the forced response, Eqs. (4.19) and (4.20) are equivalent.

Note. The dominance condition applies to the input signal. It is related to the system by the source-free solution, but it is important to distinguish conditions on the signal from conditions on the transfer function $H(s)$.

As noted several times before, in specifying a function we must describe the domain in some suitable manner. Now $F(s)$, the Laplace transform of $f(t)$, is a function whose domain consists of (complex) values of s. The function $F(s)$ can be given by a formula such as

$$F(s) = \frac{1}{s+a} \tag{4.28}$$

or by the operator of

$$F(s) = \int_0^\infty f(t)e^{-st}\,dt. \tag{4.29}$$

In either case the values of s for which the function F is defined must be specified. These values of s are determined by convergence of the integral in Eq. (4.29).

If we find a real positive finite number M such that

$$|f(t)| \leq Me^{at} \quad \text{for} \quad t \geq 0, \tag{4.30}$$

then the integral converges for all values of $s = \sigma + j\omega$, where $\sigma > a$. Thus the domain of F consists of all values of s where $\sigma > a$.

Now suppose we are given a system $H(s)$ with inverse Laplace transform $h(t)$. [In Section 4.4 we will show that $h(t)$ is the system output when the input is $\delta(t)$.] Then $H(s)$ is defined for all values of s where $\sigma > a$ [assuming, of course, that we can find an M such that $h(t)$ satisfies Eq. (4.30)]. Further, suppose the dominance condition on the input signal $X(s)$ is that $\sigma > b$. Then the output $Y(s)$ is given by

$$Y(s) = H(s)X(s), \tag{4.31}$$

where $Y(s)$ is defined for all values of s that satisfy both conditions, $\sigma > a$ and $\sigma > b$. This can be written simply as

$$\sigma > \max(a, b). \tag{4.32}$$

4.4 THE SYSTEM FUNCTION $h(t)$ IN THE TIME DOMAIN

For a stable system the source-free response will decay with time. Therefore, if Re $(s) = 0$, the dominance condition holds, and we may use $X(j\omega)$ rather than the more general $X(s)$ for the input signal. Also, we may use the transfer function $H(j\omega)$ in place of $H(s)$, since $h(t)$ can be shown to satisfy Eq. (4.30) for $a = 0$. We use this substitution in the following discussion for no particular reason other than that we dealt with Fourier transforms in the preceding chapter.

We are given a linear time-invariant system with input x and output y (see Fig. 4.7). The system is characterized by $H(j\omega)$ in the frequency domain. We know from previous experience that the output $Y(j\omega)$ is related to the input $X(j\omega)$ by multiplication. But this is in the frequency domain.

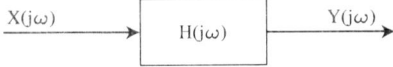

Fig. 4.7 The system $H(j\omega)$.

Suppose we are given the input $x(t)$ and wish to obtain the output $y(t)$ as a function of time. One obvious approach is to transform $x(t)$ to the frequency domain, multiply by the system function $H(j\omega)$, and transform the product back to the time domain, thus obtaining $y(t)$. Another approach is to use convolution. The convolution integral is defined by

$$y(t) = \int_{-\infty}^{+\infty} x(\lambda) h(t - \lambda)\, d\lambda$$
$$= \int_{-\infty}^{+\infty} h(\lambda) x(t - \lambda)\, d\lambda. \qquad (4.33)$$

Convolution is a binary operation, just as multiplication is. Multiplication is used in

$$Y(j\omega) = X(j\omega) H(j\omega); \qquad (4.34)$$

convolution is used in

$$y(t) = x(t) * h(t), \qquad (4.35)$$

where the symbol $*$ denotes convolution. The usefulness of this operation stems from the following two facts.

Fact 1. The transform of $H(j\omega)$ is the impulse response. That is, $H(j\omega) \leftrightarrow h(t)$, where $h(t)$ is the response to a unit impulse.

Fact 2. $X(j\omega) H(j\omega) \leftrightarrow x(t) * h(t)$.

We now illustrate the validity of these two facts.

To illustrate Fact 1, let us suppose that the input to a system is an impulse $\delta(t)$. In Fig. 4.7, $x(t) = \delta(t)$. We wish to show that the output $y(t)$ is, in this case, the transform of $H(j\omega)$. This is almost trivial* if we recognize that the

---

* Anything is trivial if you know how to do it.

4.4 THE SYSTEM FUNCTION $h(t)$ IN THE TIME DOMAIN

transform of the input is 1. That is, $\delta(t) \leftrightarrow 1 = X(j\omega)$. By Eq. (4.34),

$$Y(j\omega) = X(j\omega)H(j\omega) = H(j\omega). \qquad (4.36)$$

Therefore $y(t) = h(t)$ [taking the transform of both sides of Eq. (4.36)], and $h(t)$ is the unit impulse response.

We illustrate Fact 2 by using simple algebra. First we recognize that the Fourier transform of $x(t - \lambda)$ is given by

$$\mathscr{F}[x(t - \lambda)] = e^{-j\omega\lambda}X(j\omega), \quad \text{where} \quad x(t) \leftrightarrow X(j\omega). \qquad (4.37)$$

Now we take the Fourier transform of Eq. (4.33), and obtain

$$\begin{aligned}
\mathscr{F}[x(t) * h(t)] &= \int_{t=-\infty}^{t=+\infty} e^{-j\omega t} \left[\int_{\lambda=-\infty}^{\lambda=+\infty} x(t - \lambda)h(\lambda)\, d\lambda \right] dt \\
&= \int_{\lambda=-\infty}^{\lambda=+\infty} h(\lambda) \left[\int_{t=-\infty}^{t=+\infty} x(t - \lambda)e^{-j\omega t}\, dt \right] d\lambda \\
&= \int_{\lambda=-\infty}^{\lambda=+\infty} h(\lambda)e^{-j\omega\lambda} X(j\omega)\, d\lambda \\
&= X(j\omega)H(j\omega). \qquad (4.38)
\end{aligned}$$

We will, of course, obtain the same result if we use the convolution integral, Eq. (4.33), in the other form,

$$\int_{-\infty}^{+\infty} x(\lambda)h(t - \lambda)\, d\lambda.$$

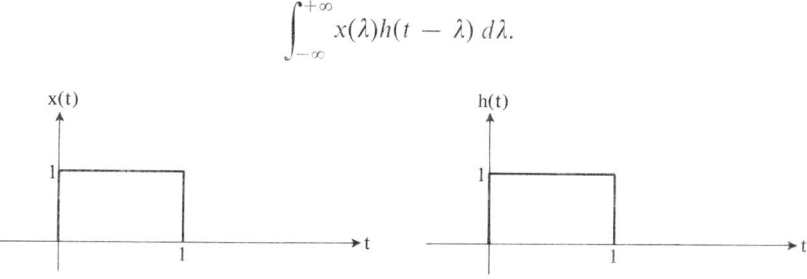

Fig. 4.8

Examples of Convolution

2. Let the input $x(t)$ and impulse response $h(t)$ be given* as in Fig. 4.8. We can determine the output, given by Eq. (4.33), *for a particular time t* by finding $h(t - \lambda)$ and $x(\lambda)$, taking the product, and then integrating over all λ. To find $x(\lambda)$, we simply substitute λ for t. The function $x(\lambda)$ is illustrated

---
* See Problem 17 for a system with this type of impulse response.

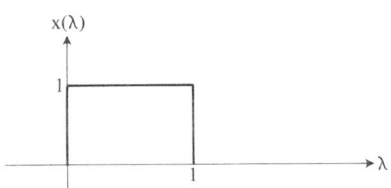

Fig. 4.9

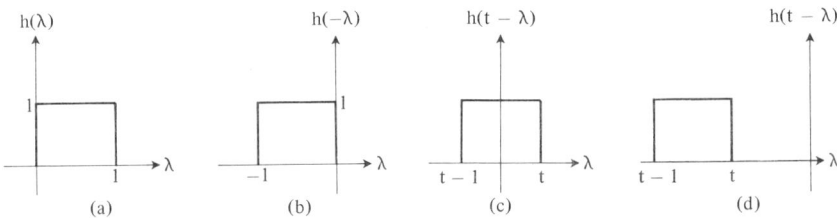

Fig. 4.10

in Fig. 4.9. Finding $h(t - \lambda)$ is slightly more complicated. First we take $h(\lambda)$, shown in Fig. 4.10(a), and "flip" it to obtain $h(-\lambda)$, Fig. 4.10(b). We then "slip" $h(-\lambda)$ to obtain $h(t - \lambda)$, Fig. 4.10(c) and (d). In Fig. 4.10(c) the value of t is 0.5. In Fig. 4.10(d) the value of t is -1. This illustrates the obvious fact that $h(t - \lambda)$ depends on t. In a similar manner, the product $x(\lambda)h(t - \lambda)$ depends on t, and therefore the integral $y(t)$ is a function of t.

We are belaboring this point because many students, on being introduced to convolution, assume that Eq. (4.33) will magically provide the entire function y rather than a single number $y(t)$. Of course, the entire function y can be generated by accounting for all possible values of time in Eq. (4.33), and this is our aim.

We continue the example. In Fig. 4.11 we give the values of time, the appropriate graphs of the product $x(\lambda)h(t - \lambda)$, and the values of $y(t)$. Thus the output $y(t)$ is given, for all t, by

$$y(t) = \begin{cases} 0, & t < 0, \\ t, & 0 < t < 1, \\ 2 - t, & 1 < t < 2, \\ 0, & t > 2. \end{cases} \quad (4.39)$$

(See Fig. 4.12.)

In this example it is obvious that $x(t) * h(t) = h(t) * x(t)$, since $x(t) = h(t)$. This is always true regardless of the shapes of $x(t)$ and $h(t)$. This can easily be shown by a change of variable in Eq. (4.33). The next example serves to illustrate this point.

4.4 THE SYSTEM FUNCTION $h(t)$ IN THE TIME DOMAIN

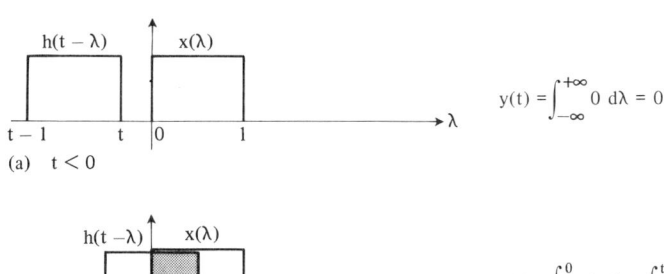

(a) $t < 0$

$$y(t) = \int_{-\infty}^{+\infty} 0 \, d\lambda = 0$$

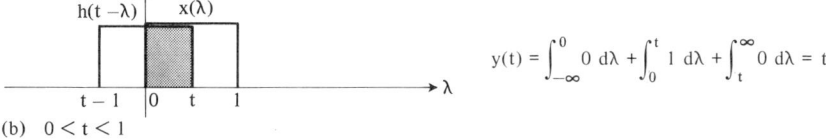

(b) $0 < t < 1$

$$y(t) = \int_{-\infty}^{0} 0 \, d\lambda + \int_{0}^{t} 1 \, d\lambda + \int_{t}^{\infty} 0 \, d\lambda = t$$

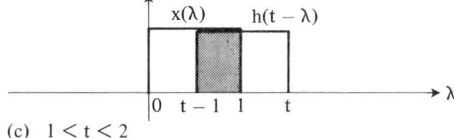

(c) $1 < t < 2$

$$y(t) = \int_{-\infty}^{t-1} 0 \, d\lambda + \int_{t-1}^{1} 1 \, d\lambda + \int_{1}^{\infty} 0 \, d\lambda = 2 - t$$

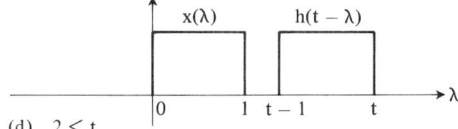

(d) $2 < t$

$$y(t) = \int_{-\infty}^{+\infty} 0 \, d\lambda = 0$$

Fig. 4.11

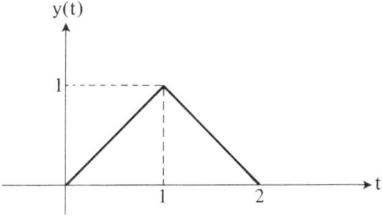

Fig. 4.12

3. Suppose that a square pulse is the input to a low-pass RC-filter (Fig. 4.13). Our first task is to find the impulse response. For this we use Fact 1, $h(t) \leftrightarrow H(j\omega)$. The system function $H(j\omega)$ is given by

$$H(j\omega) = \frac{Y(j\omega)}{X(j\omega)} = \frac{1/j\omega C}{R + 1/j\omega C}$$

$$= \frac{1/RC}{1/RC + j\omega} = \frac{a}{a + j\omega}, \quad (4.40)$$

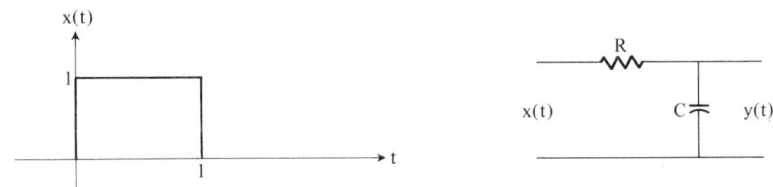

Fig. 4.13

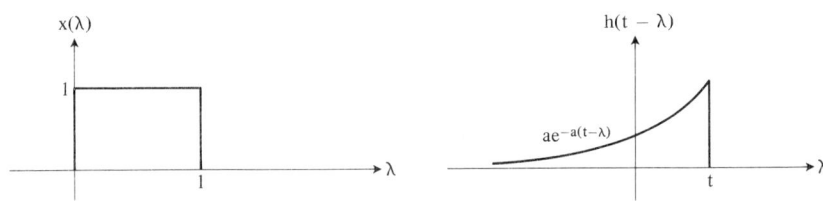

Fig. 4.14

where $a = 1/RC$. Therefore $h(t)$ is given by

$$h(t) = ae^{-at}, \quad t > 0. \tag{4.41}$$

Now to find $y(t)$ we first use Eq. (4.33) in the form

$$y(t) = \int_{-\infty}^{+\infty} x(\lambda) h(t - \lambda) \, d\lambda.$$

For this, $x(\lambda)$ and $h(t - \lambda)$ are illustrated in Fig. 4.14, where t is approximately equal to 0.5.

Figure 4.15 shows the result of convolution for various values of t, and the response curve $y(t)$.

Next we find $y(t)$ by using Eq. (4.33) in the form

$$y(t) = \int_{-\infty}^{+\infty} x(t - \lambda) h(\lambda) \, d\lambda.$$

The functions $x(t - \lambda)$ and $h(\lambda)$ are shown in Fig. 4.16 for various values of t. Note that although $x(\lambda)h(t - \lambda) \neq x(t - \lambda)h(\lambda)$, the output $y(t)$ is the same no matter which form is used. Convolution is commutive, that is,

$$x(t) * h(t) = h(t) * x(t). \tag{4.42}$$

4.5 RELATIONSHIP BETWEEN DIFFERENTIAL EQUATIONS AND THE IMPULSE RESPONSE

The differential equation characterizing the system completely describes the system. For any input $x(t)$ the output $y(t)$ can be found, provided we are able to solve the differential equation.

4.5 DIFFERENTIAL EQUATIONS AND THE IMPULSE RESPONSE

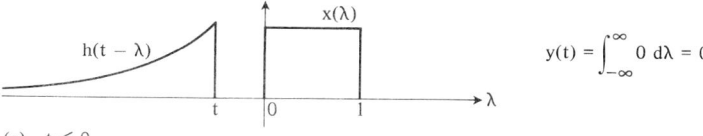

(a) $t < 0$

$$y(t) = \int_{-\infty}^{\infty} 0 \, d\lambda = 0$$

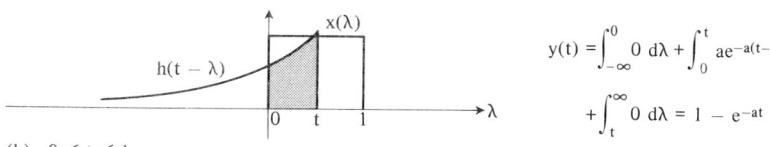

(b) $0 < t < 1$

$$y(t) = \int_{-\infty}^{0} 0 \, d\lambda + \int_{0}^{t} ae^{-a(t-\lambda)} \, d\lambda$$
$$+ \int_{t}^{\infty} 0 \, d\lambda = 1 - e^{-at}$$

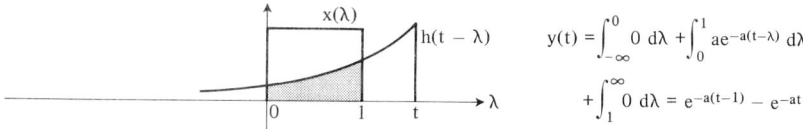

(c) $t > 1$

$$y(t) = \int_{-\infty}^{0} 0 \, d\lambda + \int_{0}^{1} ae^{-a(t-\lambda)} \, d\lambda$$
$$+ \int_{1}^{\infty} 0 \, d\lambda = e^{-a(t-1)} - e^{-at}$$

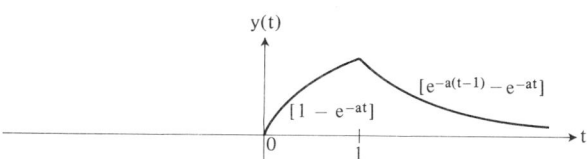

Fig. 4.15

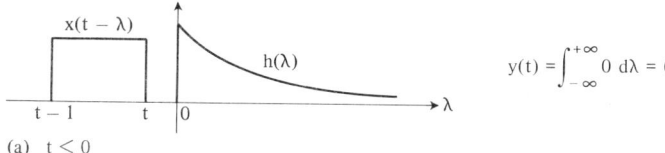

(a) $t < 0$

$$y(t) = \int_{-\infty}^{+\infty} 0 \, d\lambda = 0$$

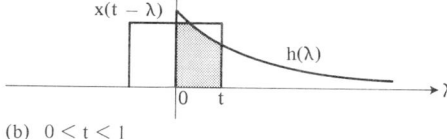

(b) $0 < t < 1$

$$y(t) = \int_{-\infty}^{0} 0 \, d\lambda + \int_{0}^{t} ae^{-a\lambda} \, d\lambda$$
$$+ \int_{t}^{\infty} 0 \, d\lambda = 1 - e^{-at}$$

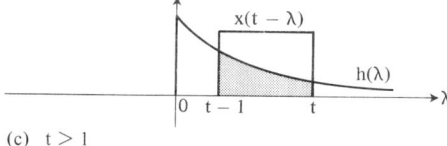

(c) $t > 1$

$$y(t) = \int_{-\infty}^{t-1} 0 \, d\lambda + \int_{t-1}^{t} ae^{-a\lambda} \, d\lambda$$
$$+ \int_{t}^{\infty} 0 \, d\lambda = e^{-a(t-1)} - e^{-at}$$

Fig. 4.16

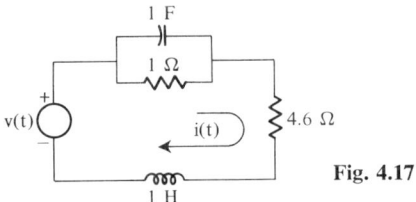

Fig. 4.17

In a similar manner, the impulse response completely describes the system. Therefore we should be able to find the impulse response if the differential equation is given. Conversely, we should be able to find the differential equation if the impulse response is given.

Finding the Impulse Response from the Differential Equation

Consider Fig. 4.17. The relationship between the current $i(t)$ and the voltage $v(t)$ is

$$\frac{dv}{dt} + v(t) = \frac{d^2i}{dt^2} + 5.6\frac{di}{dt} + 5.6\,i(t). \tag{4.43}$$

Now the impulse response is the transform of the transfer function. The transfer function $H(s)$ is given by $I(s)/V(s)$. But we can find a relationship between $I(s)$ and $V(s)$ from the differential equation (4.43). The transform of Eq. (4.43) yields

$$(s + 1)V(s) = (s^2 + 5.6s + 5.6)I(s) \tag{4.44}$$

or

$$H(s) = \frac{I(s)}{V(s)} = \frac{s + 1}{s^2 + 5.6s + 5.6}. \tag{4.45}$$

Therefore the impulse response $h(t)$ is given by

$$h(t) = \mathcal{L}^{-1}[H(s)] = 1.1e^{-4.3t} - 0.1e^{-1.3t}, \quad t \geq 0. \tag{4.46}$$

This suggests a general procedure to follow in finding the impulse response from the differential equation: Take the Laplace transform of the differential equation, and from this solve for $H(s)$, the transfer function. The inverse Laplace transform of $H(s)$ is $h(t)$.

For the general differential equation (4.12) the Laplace transform yields

$$[b_k s^k + \cdots + b_1 s + b_0]X(s) = [a_n s^n + \cdots + a_1 s + a_0]Y(s), \tag{4.47}$$

and the impulse response is the inverse Laplace transform of $Y(s)/X(s)$.

Notes. a) Our procedure is to use the Laplace transform as an intermediate step. We go from the differential equation to the transfer function $H(s)$, and then to the impulse response $h(t)$.

b) In arriving at Eq. (4.44) we assumed that there was no stored energy in the system. The presence of stored energy will alter Eq. (4.44), and therefore the impulse response $h(t)$. We will consider only the case of no stored energy.

Finding the Differential Equation from the Impulse Response

Now let us work backward and find the differential equation from the impulse response. As a specific example, consider Eq. (4.46):

$$h(t) = 1.1e^{-4.3t} - 0.1e^{-1.3t}.$$

The Laplace transform of $h(t)$ is $H(s)$ given by

$$H(s) = \frac{1.1}{s + 4.3} - \frac{0.1}{s + 1.3} = \frac{s + 1}{s^2 + 5.6s + 5.6} = \frac{I(s)}{V(s)}. \quad (4.48)$$

The input term $V(s)$ is separated from the output term $I(s)$ by the equal sign, so we have

$$(s + 1)V(s) = (s^2 + 5.6s + 5.6)I(s) \quad (4.49)$$

or

$$\frac{dv}{dt} + v(t) = \frac{d^2 i}{dt^2} + 5.6 \frac{di}{dt} + 5.6\, i(t). \quad (4.50)$$

This suggests the following general procedure in the case of no stored energy in the system:

1. Take the Laplace transform of $h(t)$. This yields $H(s)$.
2. Isolate the input term on one side of the equal sign. The resulting equation may then be transformed to give the differential equation.

PROBLEMS

1. The electric circuit shown in Fig. 4.18 can be thought of as a linear system with input $v(t)$ and output $i(t)$. For this system find
a) the differential equation that characterizes the system,
b) the transfer function $H(s)$,
c) the impulse response $h(t)$.

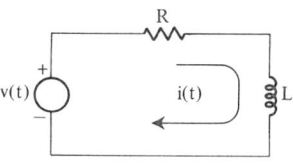

Fig. 4.18

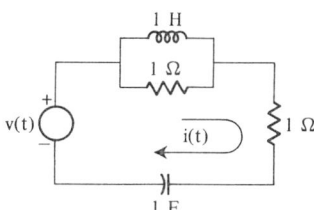

Fig. 4.19

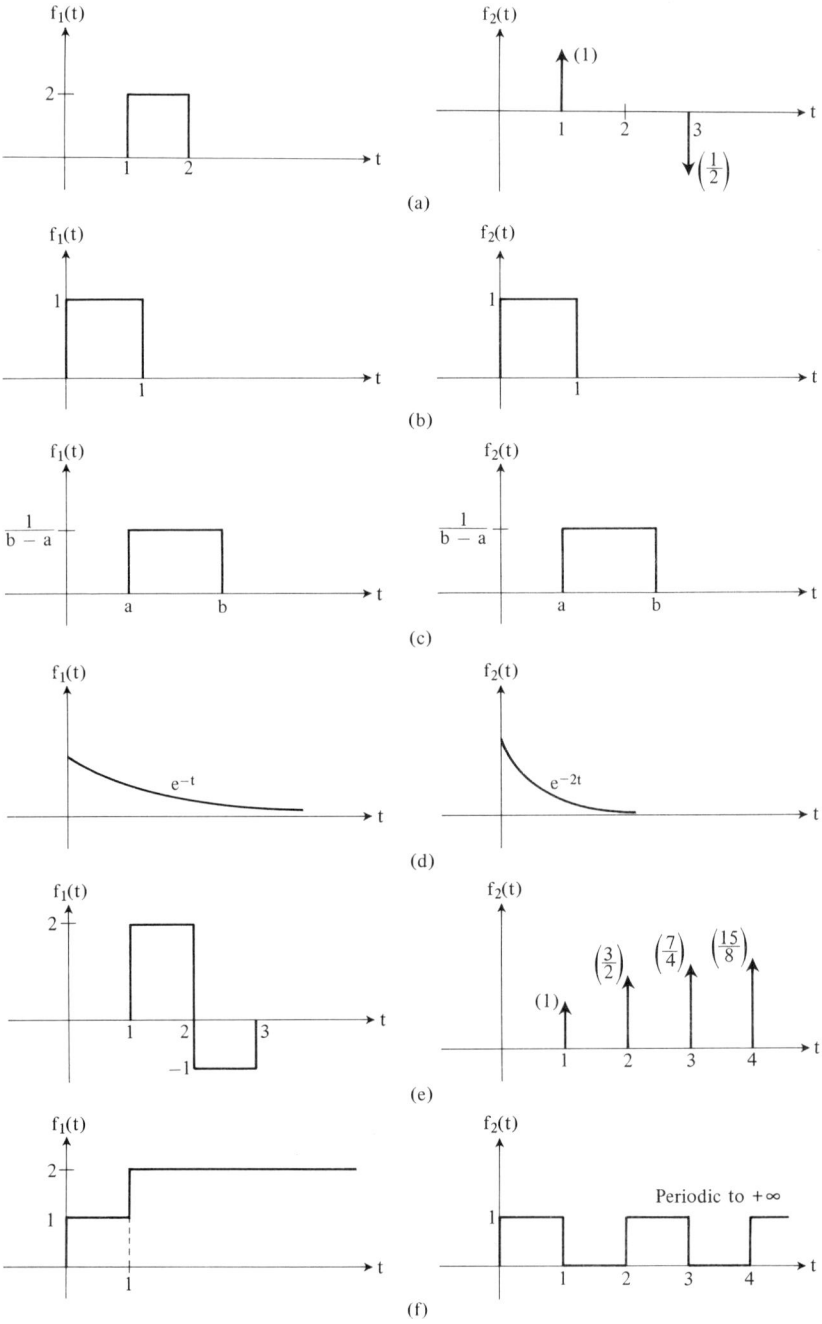

Fig. 4.20

2. Repeat parts (a), (b), and (c) of Problem 1 with the inductor L of Fig. 4.18 replaced by a capacitor C.
3. For the electric circuit shown in Fig. 4.19, where $v(t)$ is input and $i(t)$ is output, find
a) the transfer function $H(s)$,
b) the impulse response $h(t)$.
4. Convolve the function pairs shown in Fig. 4.20 and graph the results.
5. Determine the dominance condition for the circuits of Problems 1, 2, and 3. That is, for what values of s are the two definitions of $H(s)$ given in Section 4.3 equal?
6. Demonstrate the truth of the statement that convolution with a unit step function is equivalent to integration; that is,

$$f(t) * u(t) = \int_{-\infty}^{t} f(\lambda) \, d\lambda.$$

7. A "running average over the period T" is defined as

$$\frac{1}{T} \int_{t-T}^{t} f(\lambda) \, d\lambda,$$

where $f(t)$ is the time function being averaged. Express this "running average" as a convolution of $f(t)$ with another function which we will call $g(t)$. Plot and dimension $g(t)$.

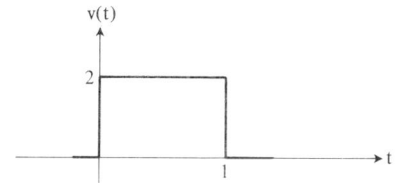

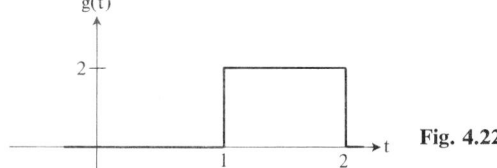

Fig. 4.21

Fig. 4.22

8. The input to the low-pass RC-filter shown in Fig. 4.21 is $v(t)$. Find the output $y(t)$.
9. Repeat Problem 8 with the input $g(t)$ shown in Fig. 4.22.
10. The input to the low-pass filter shown in Fig. 4.23 is $v(t)$. Find the output $y(t)$.

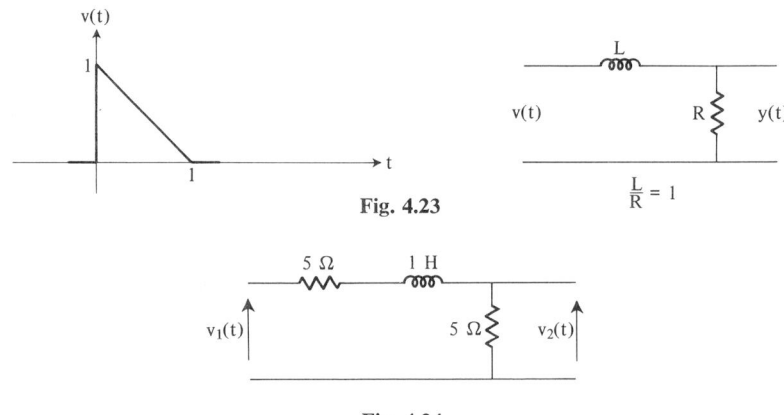

Fig. 4.23

Fig. 4.24

11. A unit impulse of voltage is applied to the input of the circuit shown in Fig. 4.24. Find $v_2(t)$.
12. a) A unit step $u(t)$ is applied to the input of Fig. 4.24. Find $v_2(t)$.
    b) Find a relationship between the unit step response and the unit impulse response. [*Hint:* Take the derivative of the answer to Problem 12(a) and compare this to the answer to Problem 11.]
13. a) A unit ramp voltage $v_1(t) = tu(t)$ is applied to the input of Fig. 4.24. Find $v_2(t)$.
    b) What is the relationship between the unit impulse, step, and ramp responses?
14. Find the response to the integral of a unit ramp, $v_1(t) = (t^2/2)u(t)$, by using your newly discovered relationship. (See Problems 11, 12, and 13.)
15. The output of the black box is given in Fig. 4.25 for input $v_1(t) = \delta(t)$. Find $v_2(t)$ if $v_1(t) = u(t)$, the unit step.

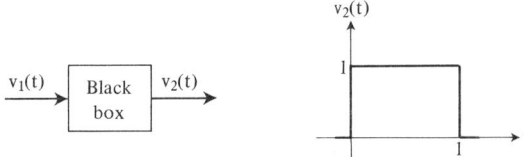

Fig. 4.25

16. Figure 4.26 shows the output of a system S for the given input, where the input is applied at $t = 0$. For this system find
    a) the free response,
    b) the forced response,
    c) the transient response,
    d) the steady-state response,
    e) the characteristic equation.

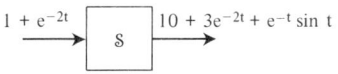

Fig. 4.26

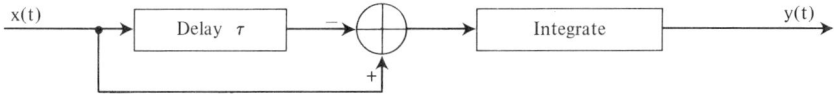

Fig 4.27

17. Find the impulse response of the system shown in Fig. 4.27.
18. The impulse response of a system is $h(t) = \frac{1}{8}[e^{-2t} - e^{-10t}]u(t)$. Find the differential equation that characterizes the system.
19. Given that the impulse response of a "black box" is

$$h(t) = \begin{cases} e^{-2t}, & t > 0, \\ 0, & t < 0, \end{cases}$$

find the step response.

20. The impulse response of a linear system with input $v(t)$ and output $i(t)$ is

$$h(t) = \frac{1}{2}[1 - e^{-4t}]u(t).$$

Find the differential equation that characterizes the system.

FURTHER READING

1. B. P. LATHI, *Signals, Systems, and Communications*, Wiley, New York, 1965.
2. R. G. BROWN and J. W. NILSSON, *Linear Systems Analysis*, Wiley, New York, 1962.
3. M. E. VAN VALKENBURG, *Network Analysis*, Prentice-Hall, Englewood Cliffs, N.J., 1964.

Each of these texts provides a general discussion of systems. Lathi has an entire chapter on convolution. Both references 2 and 3 discuss the relationship between the characteristic equation, $H(s)$, and $h(t)$. See Section 4.7 in Brown and Nilsson, and Chapter 10 in Van Valkenburg.

PART 2

Probability and Random Variables

5
PROBABILITY

5.1 SOME ELEMENTARY SET THEORY

Our study of probability will depend on an understanding of the concept of set and of set operations (such as union, intersection, complement, etc.). In this section we provide a brief outline of set theory.

Definition 1. *Set.* A set is a collection of objects that can be unambiguously classified as belonging to the collection.

Notes. a) Other names for a set are a space, a collection, a group, and a class. Often some of these terms are used for sets with special properties.

b) The items that constitute a set are called points, elements, members, and other names with similar meanings.

c) In pure mathematics "set" is usually an undefined concept. One reason for this is the difficulty of properly defining "set." For example, a thought is not a concrete object, yet a collection of thoughts is a set. By our Definition 1 it is questionable whether or not a collection of thoughts should be called a set. Although questions of this type are important, we will not be concerned with them, and Definition 1 will suffice.

d) The term "fuzzy set" was recently introduced by L. A. Zadeh. A fuzzy set is a collection of objects whose classification is ambiguous. For example, the set of all bald men is a fuzzy set because we cannot unambiguously classify a man with thin hair as either belonging or not belonging to this collection. We do not deal with fuzzy sets because Definition 1 states that it must be possible to determine membership in the set unambiguously.

Notation

We will denote sets by capital letters, $A, B, C, \ldots, X, Y, Z$. Braces { }, are also used to denote sets. For example, $A = \{a, b\}$ means that A is a set consisting of the two letters a and b.

We write $a \in A$ to denote that a is an element of the set A. We write $d \notin A$ to indicate that d is not an element of A.

There are two ways to specify a set:

1. the tabular method, whereby the elements of a set are listed within braces,
2. the rule method, whereby the elements are specified by some rule.

For example, suppose we have the set consisting of the two elements a and b. By the tabular method we specify this set by $\{a, b\}$, and by the rule method we describe it as the set consisting of the first two letters of the alphabet.

The "set-builder notation" is a combination of these two methods; it is useful when large sets are being handled. For example, it is awkward to list the elements in the set

$$B = \{1, 2, 3, 4, 5, 6, 7, 8, 9\}.$$

A more convenient notation for this set is

$$B = \{x : 0 < x < 10 \text{ and } x \text{ is an integer}\}.$$

This is read "B is the set of elements x such that x is between 0 and 10 and x is an integer." The symbol | is sometimes used in place of the symbol :. In either case, | and : are read "such that."

Definition 2. *Subset.* B is a subset of A (in symbols, $B \subset A$ or $A \supset B$) if every element of B is an element of A.

Note. Every set is a subset of itself: $A \subset A$. Thus it is not necessary for a subset to be smaller (have fewer members) than the original set. However, the subset cannot be larger (contain more members) than the original set.

Example 1. Consider the following three sets:

$$X = \{a, b, c, e, f\},$$
$$Y = \{a, b, c, d\},$$
$$Z = \{d, c, a, b\}.$$

Y is a subset of Z because every element in Y is in Z. In fact, Y and Z are equal: they are the same set. We can see that the order of the elements makes no difference. Y is not a subset of X because d is in Y but not in X. Neither is X a subset of Y. (Why?)

Definition 3. *Union.* The union of two sets A and B (in symbols, $A \cup B$) is a new set (call it C). The set C is composed of all elements in either A or B or both.

Definition 4. *Intersection.* The intersection of two sets A and B (in symbols, $A \cap B$) is a new set (call it D). The set D is composed of all elements in both A and B.

Example 2. Consider the set of all students at P.U.* Suppose that P.U., since it is a good school, teaches both mathematics and engineering.
Let

$U = \{$All P.U. students$\}$,
$A = \{$P.U. students enrolled in math$\}$,
$B = \{$P.U. students enrolled in engineering$\}$.

We can make the following observations.

1. A is a subset of U, that is, $A \subset U$. Every student in the set A is a P.U. student, and is therefore an element of U. According to Definition 2, if every element of A is also an element of U, then $A \subset U$.
2. $A \cup B$ is the set of all students enrolled in either math or engineering (or both). Note also that $(A \cup B) \subset U$.
3. $A \cap B$ is the set of students enrolled in math and engineering. That is, a student must be taking both math and engineering to be an element of $A \cap B$. Note that $(A \cap B) \subset U$.

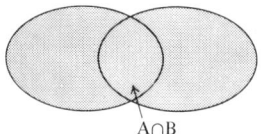

Fig. 5.1

A useful way to picture operations on sets is by Venn diagrams such as the one in Fig. 5.1. The area enclosed in a curve represents the elements of a set. Hence the total shaded area is $A \cup B$. The common, or overlapping, area is $A \cap B$. It might be helpful for the student to read the symbol $\cup$ as "or" and the symbol $\cap$ as "and." It is common practice in engineering literature to use the addition symbol for $\cup$ and the multiplication symbol for $\cap$. Then

$A \cup B = A + B$ and is read "A or B,"
$A \cap B = AB$ and is read "A and B."

We now wish to define the complement of a set. Loosely speaking, the complement of a set A is the set of all elements not in A. The complement is a special case of the difference of two sets.

Definition 5a. *Difference.* The difference $A - B$ is the set of all elements in A but not in B.

---

* This is Al Capp's P.U.—not to be confused with any existing university.

Definition 5b. *Complement.* The complement of A (written $\bar{A}$) is the difference $U - A$, where U is the universal set.

Notes. a) The universal set is the totality of all things being considered. The complement has no meaning unless the universal set has been defined (or is understood).

b) The difference of two sets is another set, just as the union and intersection of two sets are new sets. There is no correspondence between arithmetic operations such as addition, multiplication, and subtraction, and set operations.

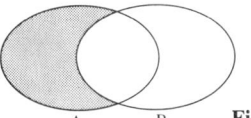

Fig. 5.2

From these definitions we see that $A - B = A\bar{B}$. That is, the difference $A - B$ is the same as the intersection of A and the complement of B. This equality is pictured in Fig. 5.2. The difference $A - B$ is represented by the shaded area (see Definition 5a). The intersection of A and $\bar{B}$ is also represented by the shaded area (see Definition 4).

We now give some examples of the above concepts.

Examples

3. Suppose we are given the sets A and B, where

$$A = \{1, 2, 3, 4\} \quad \text{and} \quad B = \{2, 4\}.$$

Then

$$A \cup B = \{1, 2, 3, 4\} = A,$$
$$A \cap B = \{2, 4\} = B,$$
$$A - B = \{1, 3\} = A\bar{B},$$
$$B - A = \varnothing.$$

The last equation says that there are no elements in the set $B - A$. The symbol $\varnothing$ denotes the *empty set*.

The complement of A has no meaning here since the universal set U has not been specified. If U is the set of integers from 1 to 10,

$$U = \{1, 2, 3, \ldots, 10\},$$

then

$$\bar{A} = \{5, 6, 7, 8, 9, 10\} = U - A.$$

4. Let the sets A and B be given by

$$A = \{1, 2, 3, 4\} \quad \text{and} \quad B = \{a, b, c\}.$$

Then
$$A \cup B = \{1, 2, 3, 4, a, b, c\},$$
$$A \cap B = \varnothing,$$
$$A - B = \{1, 2, 3, 4\} = A,$$
$$B - A = \{a, b, c\} = B.$$

Note that the intersection of A and B is the empty set; that is, A and B have no elements in common. In this case we say that A and B are *disjoint* sets.

We now consider two important related topics, equivalence and equality of sets. We encounter the concept of equivalence in counting. Basically, counting is set matching. If two sets contain the same number of elements, they can be matched; that is, each element of one set can be matched to a corresponding element in the second set.

After Polyphemus was blinded by Ulysses in the land of the Cyclops the poor old giant counted his sheep each morning by picking up a stone for each ewe that left the cave. In the evening when they returned he dropped a stone for each ewe that entered the cave.

In our present school system first graders learn to count by matching sets. They learn that a set of three blocks is equivalent to a set of three Indians, and that the number of elements in each set is three. They do this by drawing a line from the first block to the first Indian, then they draw a line from the second block to another Indian, etc., until the matching process is complete. For comparing the number of elements of two different sets, the basic idea is that of *equivalence*. Two sets A and B are equivalent if their elements can be arranged in such a way that to each element in A there corresponds exactly one element in B and to each element in B there corresponds exactly one element in A. Such a correspondence is said to be *one-to-one*, and the sets A and B are said to be equivalent. Thus we have the following definition.

Definition 6. *Equivalence.* If the elements in two sets A and B can be placed in one-to-one correspondence, the two sets A and B are said to be equivalent.

Notes. a) If for each element in A there corresponds exactly one element in B and for each element in B there corresponds exactly one element in A, then the two sets A and B are said to be equivalent.

b) This is the very idea of counting, for when we count a finite set of objects we establish a one-to-one correspondence between these objects and a set of counting numbers $\{1, 2, 3, \ldots, n\}$. A set is called finite if it is empty or if the elements can be counted, with the counting process coming to an end. This means that we can match the set of counting numbers $\{1, 2, 3, \ldots, n\}$ to the elements of a finite set, and n is the largest number needed in this matching process.

c) If a set is not finite, it is called infinite.

Equality between two sets is a stronger condition than equivalence between two sets.

Definition 7a. *Equality.* If two sets A and B are the same set, the two sets A and B are said to be equal.

This means that the two sets must contain exactly the same elements. Thus if $x \in A$, it must be true that $x \in B$. This means $A \subset B$ (see Definition 2). Also, if $x \in B$, it must be true that $x \in A$. This implies $B \subset A$. Therefore another definition of equality is the following.

Definition 7b. *Equality.* If for two sets A and B the two conditions $A \subset B$ and $B \subset A$ hold, then the two sets A and B are said to be equal.

Note. If $A = B$, then A and B are equivalent (written $A \sim B$), since their elements can be placed in one-to-one correspondence. Thus $A = B$ is a stronger or more restrictive condition than $A \sim B$.

For example, the sets $A = \{a, b, c\}$ and $B = \{b, a, c\}$ are equal. The order in which the elements are listed is immaterial. The sets $A = \{a, b, c\}$ and $C = \{1, 2, 3\}$ are not equal, although they are equivalent.

5.2 SETS AND EVENTS

Events and only events have probability. We speak of the probability of an event (that is, the probability of the occurrence of an event). The outcome of an experiment is an event. Consider the experiment of flipping a coin. There are two possible outcomes, heads and tails. The probability of heads is $\frac{1}{2}$, and the probability of tails is $\frac{1}{2}$ (provided, of course, the coin is not a trick coin).

Mathematics provides a logically consistent basis for probability theory. The concepts introduced in Section 5.1 (set theory) along with the concept of function discussed in Section 2.1 will be of use to us here. Also, it will be necessary for us to introduce some new terminology. This section is devoted to introducing these new concepts.

We wish to define "event" and "probability." Both definitions depend on the concept of sample space.

Definition 8. *Sample space.* A sample space S is the set of possible outcomes of an experiment.

Notes. a) A sample space is a *set*.

b) A sample space is the set of *all* possible outcomes of an experiment.

Examples

5. Toss a coin. The sample space consists of "heads" or "tails," that is, the sample space is the set $\{H, T\}$. Someone always asks, "But what if the coin

balances on its edge?" The answer is that we simply do not include this possibility in the sample space. If the coin does balance on its edge, the toss doesn't count and the coin must be tossed again.

6. Throw a die. There are six possible outcomes corresponding to the six die faces shown below.

$$\{ \boxdot, \boxdot, \boxdot, \boxdot, \boxdot, \boxdot \}$$

This set is the sample space. Of course, it makes writing easier if we use some symbol for the six die faces, such as $\{a_1, a_2, a_3, a_4, a_5, a_6\}$. We then call the set of these six symbols the sample space.

Definition 9. *Event.* An event is a subset of the sample space.

Notes. a) According to this definition, an event is a *set*. Any subset of the sample space can be an event. This includes the empty set $\varnothing$.

b) Since the sample space is the set of all possible outcomes, any event is a set of possible outcomes. A particular event may consist of just one experimental outcome, no experimental outcomes (the empty set), or any number of outcomes up to the total number of elements in the sample space.

A finite set with n elements has 2^n subsets. Therefore for a sample space consisting of n experimental outcomes, there are 2^n different events. In Example 5 there are four possible events: $\{H, T\}$ $\{H\}$, $\{T\}$, and $\varnothing$. When we say that the event $\{H, T\}$ occurs, we mean that *either* a head *or* a tail occurs. When we say that the event $\{H\}$ occurs, we mean that a head occurs. When we say that $\varnothing$ occurs, we mean that neither a head nor a tail occurs. Of these, $\{H, T\}$ always occurs, $\{H\}$ and $\{T\}$ occur with equal probability, and $\varnothing$ never occurs.

There are $2^6 = 64$ possible subsets of $\{a_1, a_2, a_3, a_4, a_5, a_6\}$ and therefore 64 possible events in Example 6. For instance, we might be concerned with the possibility that the number on the die is even. This is the event $\{a_2, a_4, a_6\}$, which means that either a_2, a_4, or a_6 occurs.

Definition 10. *Probability.* The probability P of an event is a function. The domain of this function is the set of all events (the set of all subsets of S), and the range is the set of numbers $0 \leq P \leq 1$. In addition, this function must satisfy three properties. For discrete sample spaces these properties are given after Definition 11 below, and for continuous sample spaces these properties are given on pages 93 and 94.

Notes. a) The range is the set of real numbers between zero and one. Therefore, the probability of an event can never be more than one or less than zero.

b) An event with probability one is called *certain*. The sample space is such an event.

c) An event with probability zero is called *null*. The empty set is a null event.

In Example 5, $P\{H, T\} = 1$, $P\{H\} = P\{T\} = \frac{1}{2}$, and $P(\varnothing) = 0$. Since an event is a set, we will use braces, { }, when the elements of the set are listed. Otherwise we will use capital letters to denote an event.

An alternative name for probability is measure. The *measure* of a set is the number assigned to that set. The function P gives the set its measure: it assigns a positive number (measure) to each subset of the sample space.

Definition 11. *Probability measure.* The probability measure is a positive number (weight) assigned to each subset (event) of the sample space so that $P(S) = 1$. (S is the sample space.)

For discrete sample spaces the probability measure has the following properties.

1. $P(X) = 0$ if and only if $X = \varnothing$.
2. $0 \leq P(X) \leq 1$ for all sets X.
3. For any two sets X and Y, $P(X \cup Y) = P(X) + P(Y)$ if and only if $X \cap Y = \varnothing$.

Definition 12. *Equiprobable measure.* An equiprobable measure is a measure that assigns the same weight to all possibilities.

For example, suppose we toss a coin twice. There are four possible outcomes, $S = \{HH, HT, TH, TT\}$, and it would be natural to assign the equiprobable measure to this sample space. We could, however, define the sample space as { 2 heads, 1 head, no heads}. We would then not want to assign a weight of $\frac{1}{3}$ to each possible outcome, since the event 1 head can occur in two different ways. Thus if S has K elements, $n(S) = K$, and if the equiprobable measure has been assigned, then each element has weight $1/K$. Therefore the measure of the subset $X \subset S$ containing r elements is r/K. That is, $P(X) = r/K$.

Finally, we must clarify the term "experiment" used in the above discussion (see Definition 8, for example). Intuitively, the term means the performance of some act with an uncertain outcome. Rolling a die and tossing a coin are examples of experiments. The possible outcomes of these experiments form the sample spaces.

In mathematical probability theory the term "experiment" has a precise meaning. An *experiment* is defined as the 3-tuple $(S, \mathfrak{F}, P)$, where S is the sample space, $\mathfrak{F}$ is the field of events (all subsets of S to which we assign probability), and P is the probability function. This is a precise way of stating the intuitive concept of performing an experiment and assigning probability to the possible outcomes. Either interpretation of the term "experiment" will be sufficient for our purposes in this text.

5.3 DISCRETE AND CONTINUOUS SAMPLE SPACES

In the above examples related to sample space and probability we have been concerned with finite sample spaces; i.e., the set of possible outcomes of an experiment has contained a finite number of elements. If we attempt to apply properties 1 and 3 of probability measure willynilly to continuous sample spaces, we encounter several difficulties. Consider the following example.

Suppose our experiment is to spin a pointer on a circle marked off with numbers from zero to one (Fig. 5.3). Now we ask, "What is the probability that the pointer will stop on the number 0.24?"

If we attempt to assign a probability measure (weight) to this outcome, we encounter a major difficulty. Suppose we decide to assign the weight $P(0.24) = 0.01$ to the outcome "pointer stops on 0.24." Since the pointer is just as likely to stop on 0.245 as on 0.24, we must also assign the weight $P(0.245) = 0.01$. In fact, we can easily select 100 numbers between 0.24 and 0.25 that the pointer can possibly stop on. And this uses up our allotment of weights!

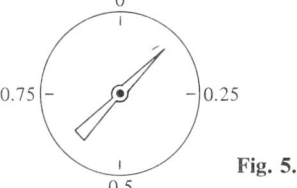

Fig. 5.3

Furthermore, no matter how small we make the measure of each outcome, we can still select enough numbers between 0.24 and 0.25 to use up the total allotment of weights.

Obviously something must be done. We can avoid this difficulty by assigning zero probability to such experimental outcomes as "the pointer stops on 0.24." Let us therefore agree to assign nonzero probability only to such outcomes as "the pointer stops *between* 0.24 and 0.25."

The set of numbers $\{x: 0.24 \leq x \leq 0.25\}$ is called measurable. Single points on a continuum have measure zero. (The symbol $\{x: 0.24 \leq x \leq 0.25\}$ is read "the set of points x, where $0.24 \leq x \leq 0.25$.")

It would be natural to assign the measure 0.01 to the interval $0.24 \leq x \leq 0.25$. Then the sum of all weights in the interval $0 \leq x \leq 1$ would be one, as required.

Recall that the probability measure (Definition 9) includes the properties

1. $P(X) = 0$ if and only if $X = \varnothing$,
3. $P(X \cup Y) = P(X) + P(Y)$ if and only if $X \cap Y = \varnothing$.

Now for continuous sample spaces, $P(X) = 0$ if the set X contains a single point. Therefore we must change property 1 to

1'. $P(S) = 1$.

Also, in property 3 it is possible for $X \cap Y$ to contain a single point. Therefore property 3 becomes

3'. $P(X \cup Y) = P(X) + P(Y)$ if $X \cap Y = \varnothing$.

In conclusion, we may use the properties given after Definition 11 for finite or discrete sample spaces. For continuous sample spaces, properties 1 and 3 must be changed as above.

5.4 PARTITIONING

The concept of the partition of a set is both useful and necessary for our subsequent study of probability. A *partition* of a set is a subdivision of the set into smaller sets (called *cells*) so that all these smaller sets may be combined to form the original set, with nothing left over and nothing left out. A precise definition is as follows:

> **Definition 13.** *Partition.* A partition of a set U is a subdivision of U into subsets that are *disjoint* and *exhaustive*. Thus $[A_1, A_2, \ldots, A_n]$ is a partition of U if
>
> 1. $A_i \cap A_j = \varnothing$, $i \neq j$ (disjoint),
> 2. $A_1 \cup A_2 \cup \cdots \cup A_n = U$ (exhaustive).

Notes. a) Statement 1 is the definition of "disjoint" (remember that $\varnothing$ is the null or empty set). This means that any element of U can be in at most one cell.

b) Statement 2 is the definition of "exhaustive." The cells $A_1, A_2, \ldots, A_n$ exhaust U, which means that every element of U is in at least one cell.

c) The basic idea of a partition is that each element of U is in exactly one cell. Another definition of partition is: A division of U into cells such that each element of U is in exactly one cell is said to be a partition of U.

Examples

7. Suppose U is the set of all students in a given classroom. Then A_1 and A_2 form a partition of U if

$$A_1 = \text{the set of all men in the class,}$$
$$A_2 = \text{the set of all women in the class.}$$

The property of a partition that makes it so useful is that the total number of elements in U is just the sum of all the elements in each of the cells that combine to form the partition. That is,

$$n(U) = n(A_1) + n(A_2) + \cdots + n(A_n), \tag{5.1}$$

where $n(X)$ denotes the number of elements in the set X. This property is a result of the fact that each element of U is in one and only one of the cells $A_1, A_2, \ldots, A_n$.

8. Given two sets A and B, find the number of elements in $A \cup B$ in terms of $n(A)$, $n(B)$, and $n(AB)$.

The set A can be partitioned into disjoint sets $A \cap B$ and $A \cap \bar{B}$. Also B can be partitioned into $A \cap B$ and $\bar{A} \cap B$. Then

$$n(A) = n(A \cap B) + n(A \cap \bar{B}), \tag{5.2}$$

$$n(B) = n(A \cap B) + n(\bar{A} \cap B). \tag{5.3}$$

Adding, we get

$$n(A) + n(B) = 2n(A \cap B) + n(A \cap \bar{B}) + n(\bar{A} \cap B). \tag{5.4}$$

Since $(A \cap B)$, $(A \cap \bar{B})$, and $(\bar{A} \cap B)$ are disjoint and their union is $A \cup B$, they form a partition of $A \cup B$. Then

$$\begin{aligned} n(A \cup B) &= n(A \cap B) + n(A \cap \bar{B}) + n(\bar{A} \cap B) \\ &= n(A) + n(B) - n(A \cap B). \end{aligned} \tag{5.5}$$

The last term in Eq. (5.5) is obtained from Eq. (5.4). It is obvious from the Venn diagram in Fig. 5.4 that Eq. (5.5) is correct, for when we add $n(A) + n(B)$, those elements common to both A and B have been added twice. The elements common to both A and B form the set $A \cap B$, so we must subtract the number of elements in $A \cap B$ to arrive at $n(A \cup B)$.

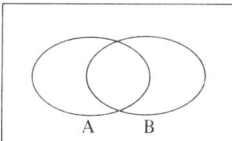

Fig. 5.4

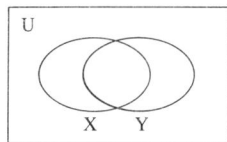

Fig. 5.5

5.5 CONDITIONAL PROBABILITY

The universal set U must be specified before measure can be assigned to subsets of U. Suppose X and Y are subsets of U (Fig. 5.5), with the number of elements in each set given by

$$n(X) = 10, \tag{5.6}$$

$$n(Y) = 20, \tag{5.7}$$

$$n(U) = 100. \tag{5.8}$$

Then, using the equiprobable measure, we let

$$P(X) = \tfrac{10}{100} = 0.1, \tag{5.9}$$

$$P(Y) = \tfrac{20}{100} = 0.2. \tag{5.10}$$

However, if $n(U) = 50$, say, then we double $P(X)$ and $P(Y)$. Therefore the measure assigned to subsets of U depends on U.

Conditional probability is in effect a change of the universal set. We will use an example to introduce this concept.

Let U be all students at P.U. (Using the above figures, P.U. is a small school with a total enrollment of 100.) Let X be the set of students enrolled in engineering, and let Y be the set of students enrolled in math.

Suppose our experiment is the selection of a P.U. student at random. (The phrase "at random" means that we are just as likely to select one student as another.) Then U is the sample space, and the sets X and Y are events:

X is the event "The student is taking engineering."
Y is the event "The student is taking math."

Now suppose that five P.U. students are enrolled in both math and engineering. Then
$$n(X \cap Y) = 5, \tag{5.11}$$
so that
$$P(X \cap Y) = \tfrac{5}{100} = 0.05. \tag{5.12}$$

We wish to find the conditional probability of X, given Y. In symbols this is denoted by $P(X|Y)$. That is, we are given the information that the selected student is enrolled in math (belongs to the set Y) and we wish to find the probability that this student is also enrolled in engineering (also belongs to the set X).

Obviously, the universal set has been changed. No longer is the selected student just any of the 100 P.U. students. He is one of the 20 students enrolled in math. Thus our new universal set becomes Y. Figure 5.6 illustrates the new universal set Y, along with the subset $X \cap Y$. From the diagram we see that
$$P(X|Y) = \frac{P(X \cap Y)}{P(Y)} = \frac{0.05}{0.2} = 0.25. \tag{5.13}$$

Similar reasoning can be used to show that
$$P(Y|X) = \frac{P(X \cap Y)}{P(X)}. \tag{5.14}$$

Solving Eqs. (5.13) and (5.14) for $P(X \cap Y)$, we arrive at the fundamental relationship
$$P(X \cap Y) = P(X|Y)P(Y) \tag{5.15}$$
$$= P(Y|X)P(X). \tag{5.16}$$

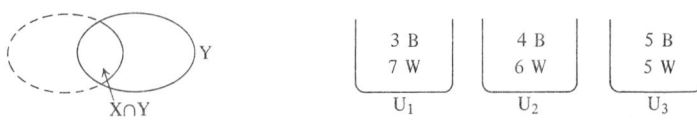

Fig. 5.6   Fig. 5.7

Bayes' Law

Bayes' law is derived from Eqs. (5.15) and (5.16). Bayes' formula, law, or rule states that

$$P(X|Y) = \frac{P(Y|X)P(X)}{P(Y)}. \tag{5.17}$$

The following example illustrates some features of Bayes' rule.

Three urns, U_1, U_2, and U_3, contain black marbles and white marbles as shown in Fig. 5.7. An urn is chosen at random and a marble is drawn from this urn. The marble is black. What is the probability that it came from U_3?

We define the following events:

B = the marble drawn is black,
U_1 = the marble came from U_1,
U_2 = the marble came from U_2,
U_3 = the marble came from U_3.

Alternatively, we may think of these events as subsets of the universal set U, where U is the set of all marbles in the three urns. Then B is the set of black marbles, U_1 is the set of marbles in urn 1, etc.

The problem asks for $P(U_3|B)$. By Eq. (5.17) this is given by

$$P(U_3|B) = \frac{P(B|U_3)P(U_3)}{P(B)}. \tag{5.18}$$

We can easily compute the numerator of Eq. (5.18), as follows:

$$P(B|U_3) = \tfrac{1}{2}, \tag{5.19}$$

$$P(U_3) = \tfrac{1}{3}. \tag{5.20}$$

We now contend that $P(B)$ is given by

$$P(B) = P(B|U_1)P(U_1) + P(B|U_2)P(U_2) + P(B|U_3)P(U_3)$$
$$= (\tfrac{3}{10})(\tfrac{1}{3}) + (\tfrac{4}{10})(\tfrac{1}{3}) + (\tfrac{5}{10})(\tfrac{1}{3}) = 0.4, \tag{5.21}$$

so that the solution to our problem is

$$P(U_3|B) = \frac{(\tfrac{1}{2})(\tfrac{1}{3})}{0.4} = \frac{1}{2.4}. \tag{5.22}$$

Derivation of Eq. (5.21). We want to exhibit the conditions that must hold for Eq. (5.21) to be true.

Suppose B can occur only if U_1, U_2, or U_3 occurs. Also suppose the U_i's are mutually exclusive, that is, $U_i \cap U_j = \emptyset$, $i \neq j$. Now B can be partitioned (Definition 13) into disjoint sets $(B \cap U_1)$, $(B \cap U_2)$, and $(B \cap U_3)$. (See Fig. 5.8.) This means that

$$B = BU_1 \cup BU_2 \cup BU_3. \tag{5.23}$$

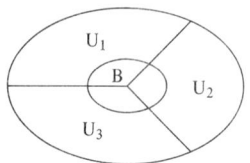

Fig. 5.8

By property 3 of probability measure we have

$$P(B) = P(BU_1) + P(BU_2) + P(BU_3), \tag{5.24}$$

and from Eq. (5.15) this gives the desired result, Eq. (5.21).

Bayes' formula is most useful in the form

$$P(U_3|B) = \frac{P(B|U_3)P(U_3)}{\Sigma_i P(B|U_i)P(U_i)}, \tag{5.25}$$

which is valid so long as B can occur only if one of the U_i's occurs, and where the U_i's are mutually exclusive.

5.6 MUTUALLY EXCLUSIVE AND INDEPENDENT EVENTS

We cannot always depend on intuition in matters based on logic. Intuitively, two events are mutually exclusive if the occurrence of one prohibits the occurrence of the other. Mathematically, we define the concept of mutually exclusive events as follows:

Definition 14. *Mutually exclusive events.* Two events A and B are mutually exclusive if and only if

$$P(A \cup B) = P(A) + P(B). \tag{5.26}$$

Note. This definition coincides with our intuitive definition, for it is always true that

$$P(A \cup B) = P(A) + P(B) - P(A \cap B). \tag{5.27}$$

(Refer to property 3 of probability measure.) If $P(A \cap B)$ is zero, then the two events are nonoverlapping (Fig. 5.9b) and $P(A \cup B)$ is given by Eq. (5.26).

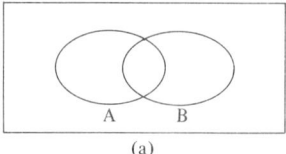

(a)

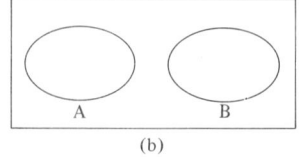

(b)

Fig. 5.9 (a) A and B not mutually exclusive. (b) A and B mutually exclusive.

5.6 MUTUALLY EXCLUSIVE AND INDEPENDENT EVENTS

Example 9. Consider an ordinary deck of playing cards. Define the events A, B, and C as

$$A = \text{an ace is selected,}$$
$$B = \text{a heart is selected,}$$
$$C = \text{a club is selected.}$$

Clearly A and B are not mutually exclusive. Therefore

$$P(A \cup B) = P(A) + P(B) - P(A \cap B)$$
$$= \tfrac{1}{13} + \tfrac{1}{4} - \tfrac{1}{52} = \tfrac{16}{52}.$$

However, B and C are mutually exclusive. Therefore

$$P(B \cup C) = P(B) + P(C)$$
$$= \tfrac{1}{4} + \tfrac{1}{4} = \tfrac{1}{2}.$$

Another important property of events is independence. Intuitively, *independence* means that the occurrence of one event does not affect the occurrence or nonoccurrence of another event. Then two events A and B are independent if the probability of A given that B has occurred is exactly equal to the probability of A; that is,

$$P(A|B) = P(A). \tag{5.28}$$

But, by Eq. (5.15), it is always true that

$$P(A \cap B) = P(A|B)P(B)$$
$$= P(B|A)P(A). \tag{5.29}$$

We can now define independence.

Definition 15. *Independence.* Two events A and B are (statistically) independent if and only if

$$P(A \cap B) = P(A)P(B). \tag{5.30}$$

Notes. a) In view of Eq. (5.29), this is equivalent to Eq. (5.28). Either condition can be taken as the definition of statistical independence.

b) Note the emphasis on "statistical" independence. There is also "functional" independence, which is something quite different from statistical independence.

Example 10. Let us define A, B, and C as in Example 9. Consider

$$P(A \cap B) = P(A|B)P(B),$$

where

$$P(A|B) = \tfrac{1}{13}, \quad \text{(There is one ace out of 13 hearts.)}$$
$$P(B) = \tfrac{1}{4}.$$

Therefore

$$P(A \cap B) = \tfrac{1}{13} \cdot \tfrac{1}{4} = \tfrac{1}{52}$$

but
$$P(A) = \tfrac{4}{52} = \tfrac{1}{13} = P(A|B).$$

We conclude that the events A and B are statistically independent.

Now consider events B and C. We find that

$P(B|C) = 0$, (It is impossible to select a heart and a club on the same draw.)

$P(B) = \tfrac{1}{4}$.

Therefore B and C are not statistically independent.

Events cannot be both independent and mutually exclusive, so long as their probabilities are not zero. To prove this statement we will use the method of contradiction; that is, we assume that events A and B are both mutually exclusive and independent, and then show that a contradiction results.

Let events A and B have nonzero probability:

$$P(A) > 0, \qquad (5.31)$$
$$P(B) > 0. \qquad (5.32)$$

Assume that A and B are mutually exclusive. By Definition 12 this means that

$$P(A \cap B) = 0. \qquad (5.33)$$

Now assume that A and B are also independent. By Definition 13 this means that

$$P(A \cap B) = P(A)P(B). \qquad (5.34)$$

But by Eqs. (5.31) and (5.32) the product $P(A)P(B)$ cannot equal zero. Thus we have a contradiction, and our statement is proved.

5.7 EXAMPLES

Solving probability problems is something like evaluating integrals: it's anybody's game, and there are usually several approaches to each problem. However, there are certain steps that one can follow in solving typical problems. This procedure is especially useful for the novice, at least until he has had some practice at the art. The procedure is outlined in four steps as follows:

1. Sketch the sample space. That is, draw a picture that somehow illustrates the possible outcomes of the experiment.
2. Define the appropriate events.
3. Write down the probability of all these events that you can.
4. Determine what is needed to complete the solution.

The examples that follow illustrate this procedure.

5.7 EXAMPLES

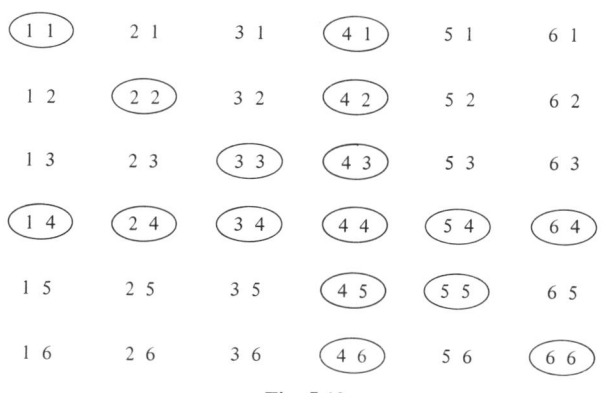

Fig. 5.10

Examples

11. Three dice are thrown. Determine the probability that at least two of them will turn up the same face.

Step 1 is to sketch the sample space. There are $6^3 = 216$ different possible experimental outcomes, and therefore it would take much time (and paper) to list them. We can simplify our problem by listing only the $6^2 = 36$ possible outcomes for the first and second dice. This list is shown in Fig. 5.10.

Now the problem can be solved regardless of which face comes up on the third die. Suppose 4 turns up on the third die. Then there are 16 possible ways for at least two of the dice to turn up the same face. These 16 ways are circled in Fig. 5.10. Since there are 36 possibilities, the probability of at least two dice having the same face is $\frac{16}{36} = \frac{4}{9}$.

Note. We solved this problem using little more than step 1 of our procedure. We now solve a slightly more complicated problem involving Bayes' formula.

Fig. 5.11

12. Suppose we have three boxes with gold and silver coins distributed as shown in Fig. 5.11. Select a box at random and select a coin from this box. If the coin is gold, what is the probability that there is another gold coin in the box?

The figure serves as a sketch of the sample space (step 1). For step 2, let us define the following events:

S is the event "A silver coin is selected."
G is the event "A gold coin is selected."
B_1 is the event "Box 1 is selected."
And so on for B_2 and B_3.

We can write down the probability of the events B_1, B_2, and B_3 (step 3):

$$P(B_1) = P(B_2) = P(B_3) = \tfrac{1}{3}.$$

Now the problem statement asks for $P(B_2|G)$. By Bayes' formula this is

$$P(B_2|G) = \frac{P(G|B_2)P(B_2)}{\sum_{i=1}^{3} P(G|B_i)P(B_i)}$$

$$= \frac{\tfrac{2}{5}\tfrac{1}{3}}{\tfrac{2}{5}\tfrac{1}{3} + \tfrac{1}{3}\tfrac{1}{3} + \tfrac{1}{2}\tfrac{1}{3}} = \frac{12}{37}.$$

13. Assume that in the World Series each team has probability $\tfrac{1}{2}$ of winning each game, regardless of the outcomes of any other games. Suppose team A has already won three games and team B has won only one. Compute the probability of team A winning the series. (The World Series is won by the first team to win four games.)

Fig. 5.12   Fig. 5.13

Figure 5.12 shows a sketch of the sample space.
Let us define the following events:

A is the event "Team A wins the series."
G_5 is the event "Team A wins the series on game 5."
W_5 is the event "Team A wins game 5."
L_5 is the event "Team A loses game 5."
And so on for G_6, W_6, ...

The event A occurs if G_5 or G_6 or G_7 occurs. That is, $A = G_5 \cup G_6 \cup G_7$, where G_5, G_6, and G_7 are mutually exclusive. Therefore.

$$P(A) = P(W_5) + P(W_6|L_5)P(L_5) + P(W_7|L_5, L_6)P(L_6|L_5)P(L_5) = \tfrac{7}{8}.$$

PROBLEMS

1. Sets A, B, and C have the relationship shown in the Venn diagram in Fig. 5.13.
   a) Indicate the set $A \cap B$ by shading.
   b) Shade $A \cap B \cap C$.
   c) Shade $A \cap B \cap \bar{C}$.
   d) Shade $\overline{A \cap B \cap C}$.

2. The two sets A and B are given by
$$A = \{-5, -4, -3, -2, 0, 1, 2, 3, 4, 5\},$$
$$B = \{1, 2, 3, 4, 5\}.$$
a) Find $A - B$.
b) Find $B - A$.
c) Find $A \cup B$.
d) Find $A \cap B$.

3. Let A be the set of numbers between 0 and 1. In symbols, $A = \{x : 0 < x < 1\}$. This is read "A is the set of numbers x such that x is between zero and one." Also, the sets B and C are defined as
$$B = \{x : -1 < x < 1\}, \qquad C = \{x : -1 < x < \tfrac{1}{2}\}.$$
a) Is $A \subset B$, $A \subset C$, $C \subset B$, $A \subset A$?
b) Find $A \cup B$, $A \cup C$, $A \cap B$, $A \cap C$, $B - A$, $A - B$.

4. Let the sample space S consist of the 10 letters $a, b, c, \ldots, j$, with each letter equally likely to occur. Define the events A, B, and C as
$$A = \{a, b, c, d\}, \qquad B = \{a, c\}, \qquad C = \{d, e, f\}.$$

Calculate the probabilities of the following events:

a) $A \cup B$,   b) $A \cup \bar{B}$,   c) $\overline{A \cup B}$,   d) $\bar{A} \cap \bar{B}$,
e) $A \cap B \cap C$,   f) $A \cap (B \cup C)$,   g) $(A \cap B) \cup C$.

5. A single die with faces marked 1 through 6 is thrown. Define events A, B, and C as follows:

A is the event "The number is even."
B is the event "The number is larger than 3."
C is the event "The face 2 turns up."

a) What is the sample space?

Calculate the probabilities of the following events:

b) $A \cup B$,   c) $A \cap B$,   d) $B \cap C$,   e) $A \cap \bar{C}$.

6. A pointer is spun on a circle marked off from 0 to 1 (see Fig. 5.14). Calculate the probabilities of the following events:
a) The pointer stops between 0 and 0.25.
b) The pointer stops between 0 and 0.5.
c) The pointer stops between 0 and 0.75.
d) The pointer stops on 0.5.

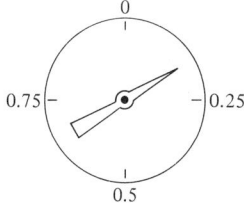
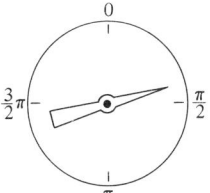

Fig. 5.14   Fig. 5.15

7. The circle is now marked off from 0 to 2π (see Fig. 5.15).
   a) What is the sample space?
   b) Calculate the probability that the pointer stops between 0 and π.
   c) Calculate the probability that the pointer stops between 0 and 3.
   d) Given that the pointer stopped between 0 and 3, what is the probability that it stopped between 0 and $\pi/2$?

8. Let A, B, and C be any three sets. Find the number of elements in the union of A, B, and C, $n(A \cup B \cup C)$, in terms of $n(A)$, $n(B)$, $n(C)$, $n(AB)$, $n(AC)$, $n(BC)$, and $n(ABC)$.

9. Let X be an event. Show that
$$P(\bar{X}) = 1 - P(X),$$
where $\bar{X}$ denotes the complement of X.

10. What is the meaning of the statement "The events A and B are mutually exclusive"?

11. Two events $A = \{a, b, c, d, e, f, g\}$ and $B = \{e, f, g, h, i\}$ are independent. What are the elements of
    a) $A \cup B$?   b) $A \cap B$?

12. A simple game with four cards is played as follows. The cards, labeled 1, 2, 3, and 4, are shuffled and dealt face up one at a time. You win the game whenever you are dealt a card whose value exceeds the value of the first card dealt.
    a) Define the sample space for this experiment.
    b) Calculate the probability of winning on the second deal.
    c) Calculate the probability of losing.

13. A drunk staggers out of a bar at the corner of Oak and Elm Streets. He is so befuddled that he cannot tell direction, but he knows from long experience that his home is exactly four blocks from the bar. Suppose he has an equal chance of traveling in any of the four directions when he reaches any corner (note that he is just as likely to retrace his steps as to go in any of the other three directions). Given that he travels exactly four blocks, what is the probability that he will end up back at the corner of Oak and Elm?

14. Find the probability of hitting a ship if one fires four torpedoes, each having a probability of $\frac{1}{4}$ of scoring a hit.

15. Three men, A, B, and C, each fire one shot at a target. The probabilities of their hitting the target are
$$P(A) = 0.30, \quad P(B) = 0.25, \quad P(C) = 0.10.$$
Compute
    a) the probability of at least one hit,
    b) the probability of no hits,
    c) the probability that the one bullet found in the target came from A's gun.

16. There are three urns containing red and black balls as shown in Fig. 5.16. An urn is chosen, a ball drawn, and the color noted. It is red. What is the probability that it came from urn 1?

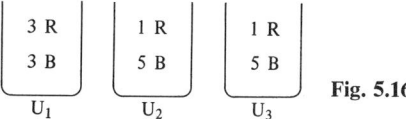

Fig. 5.16

17. Many students remain unconvinced of the validity of Bayes' law. With regard to Problem 16, they would contend that the probability of choosing urn 1 is $\frac{1}{3}$ regardless of the color of the ball drawn. For those who remain in doubt, let us propose the following:

As in Problem 16, an urn is chosen, a ball drawn, and the color noted. After seeing the ball you must state whether or not you believe the ball came from urn 1.

Note. You have only one of two decisions to make:
1) Yes—the ball came from urn 1.
2) No—the ball did not come from urn 1.

Answer the following questions:
a) What is your decision rule? That is, if the ball is red, what is your decision? If the ball is black, what is your decision?
b) If the ball is black, what is the probability that you will make an error, that is, what is $P(e|B)$?
c) If the ball is red, what is $P(e|R)$?
d) Compute the probability of error, $P(e) = P(e|B)P(B) + P(e|R)P(R)$.

Note. If knowledge of the color of the ball had no effect on the probability that urn 1 was chosen, then your decision rule would be to say, "No, this ball did not come from urn 1," regardless of the color of the ball, and you would be right $\frac{2}{3}$ of the time (that is, $P(e) = \frac{1}{3}$). Is your decision rule better than this?

18. *Bertrand's box paradox.* Suppose we have three boxes each containing two coins. In box 1 there are two gold coins; in box 2 there are two silver coins; and in box 3 there is a gold coin and a silver coin. A box is selected at random and a coin withdrawn. It is gold. What is the probability that the other coin in this same box is silver?

19. a) Find the probability of getting a 7 in the throw of a pair of dice.
    b) Find the probability of getting either a 6, 7, or 8 in the throw of a pair of dice.

20. A box contains five white balls, three red balls, and two black ones. What is the probability that two balls drawn from the box will both be red?

21. a) What is the probability of obtaining four tails if four coins are tossed?
    b) What is the probability that at least three heads will appear?

22. The game of craps consists of a "shooter" rolling two dice and someone else betting against him. The shooter wins outright if the dice total 7 or 11, and loses outright if they total 2, 3, or 12. When 4, 5, 6, 8, 9, or 10 occurs on the first roll, then this number becomes the shooter's "point," and he continues to roll until either his point or 7 comes up. If the former is first, the shooter wins; if the latter is first, he loses. What is the probability of the shooter winning?

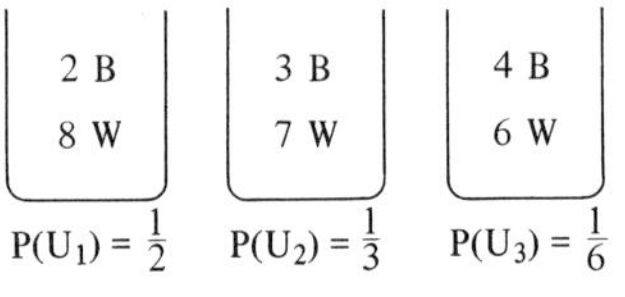

$$\begin{array}{ccc} \boxed{\begin{array}{c} 2\text{ B} \\ 8\text{ W} \end{array}} & \boxed{\begin{array}{c} 3\text{ B} \\ 7\text{ W} \end{array}} & \boxed{\begin{array}{c} 4\text{ B} \\ 6\text{ W} \end{array}} \\ P(U_1) = \tfrac{1}{2} & P(U_2) = \tfrac{1}{3} & P(U_3) = \tfrac{1}{6} \end{array}$$

Fig. 5.17

23. I have three urns behind a curtain. These urns are arranged so that urn 1 is easier to reach than urn 2, which in turn is easier to reach than urn 3. Black and white marbles are distributed in the urns as shown in Fig. 5.17. Suppose you reach behind the curtain, fumble around for a marble, find one, and show it to me. It is black. What is the probability that it came from urn 1? The probabilities of choosing urns 1, 2, and 3 are shown in the figure.

FURTHER READING

1. M. McFadden, *Sets, Relations, and Functions*, McGraw-Hill, New York, 1963.
2. P. L. Meyer, *Introductory Probability and Statistical Applications*, Addison-Wesley, Reading, Mass., 1965.
3. A. Papoulis, *Probability, Random Variables, and Stochastic Processes*, McGraw-Hill, New York, 1965.
4. J. G. Kemeny, H. Mirkil, J. L. Snell, and G. L. Thompson, *Finite Mathematical Structures*, Prentice-Hall, Englewood Cliffs, N.J., 1958.

Set theory is discussed in most elementary math texts, but McFadden provides a thorough introduction to the subject. References 2, 3, and 4 each provide an excellent introduction to probability, and reference 4 provides additional material on set theory.

6
RANDOM VARIABLES

6.1 DEFINITION OF RANDOM VARIABLE

The term "random variable" is a classic misnomer, for it totally fails to describe a random variable. A random variable is neither random nor a variable. A random variable is a function.

> **Definition 1.** *Random variable.* A random variable is a function with domain S (the sample space). In addition, this function must satisfy the following two properties:
> 1. The set $\{X \leq x\}$ is an event for every real number x.
> 2. $P\{X = \infty\} = 0$, $P\{X = -\infty\} = 0$.

Notes. a) Although we have not specified the range in Definition 1, we will deal only with real-valued random variables, that is, with random variables with range the set of real numbers.

b) Probability is also a function whose domain is related to the sample space (see Definition 10, Chapter 5). There are several important differences: (1) The domain for random variables is just S, the sample space. The domain for probability is all subsets of S (all events). Hence, their domains are different. (2) The range for a random variable can be any set of real numbers. (3) There is an important conceptual difference. Probability is assigned before the experiment is performed. (After all, if we knew the experimental outcome, we would assign probability one to any event that included the actual outcome and probability zero to all other events.) A random variable is a function that assigns numbers to the experimental outcomes, and it is immaterial whether the experiment has been performed or not.

It is conventional to use a capital letter (usually X, Y, or Z) to denote a random variable, and the corresponding lower-case letter (x, y, or z) to denote an element in the range, i.e., a particular value of the random variable. A Greek letter is used to denote an element in the domain (a possible experimental outcome). Some examples will illustrate these concepts.

Examples

1. Suppose you match coins with a friend, winning $1.00 if the two coins match and losing $1.00 if the coins do not match. A random variable may be defined as the mapping from the sample space $\{HH, HT, TH, TT\}$ to the numbers $+1$ and -1. We identify the elements in the domain with Greek letters ζ_i as follows:

$$\zeta_1 = HH, \quad \zeta_2 = HT, \quad \zeta_3 = TH, \quad \zeta_4 = TT. \tag{6.1}$$

If our random variable is labeled X, then

$$X(\zeta_1) = 1, \quad X(\zeta_2) = -1, \quad X(\zeta_3) = -1, \quad X(\zeta_4) = 1. \tag{6.2}$$

There are four events in the domain and only two numbers in the range.

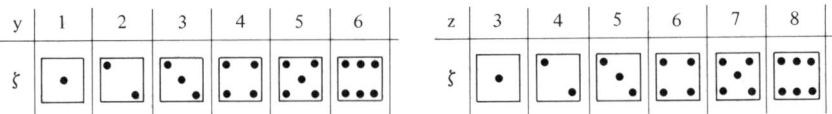

Fig. 6.1 The random variable Y. **Fig. 6.2** The random variable Z.

2. Figure 6.1 illustrates the random variable Y. A single die is thrown. The random variable Y assigns the numbers 1, 2, 3, 4, 5, and 6 to the six possible outcomes ζ_i.

Figure 6.2 illustrates another random variable, Z. The random variable Z is different from Y, but is a perfectly good random variable nevertheless.

3. An audio amplifier is amplifying music. At time t_1 we measure the voltage at the output terminals of the amplifier. This voltage $v(t_1)$ is a number in the range of our random variable. The domain consists of all possible potential differences at the output terminals. (In effect, the voltmeter assigns the number $v(t_1)$ to the experimental outcome "potential difference." One is tempted to call the voltmeter the random variable. Of course, a voltmeter is a voltmeter, not a function.)

6.2 DISTRIBUTION OF RANDOM VARIABLES

We now wish to introduce two functions: the cumulative distribution function (cdf) and the probability density function (pdf).

Definition 2. *Cumulative distribution function (cdf).* $F_X(x)$ is the cdf of the random variable X. It is defined by

$$F_X(x) = P\{X \leq x\}. \tag{6.3}$$

6.2 DISTRIBUTION OF RANDOM VARIABLES

Notes. a) By the symbol $\{X \leq x\}$ we mean the event "The value of the random variable X is less than or equal to x," where x is just some number.

b) The cdf is therefore a function of x. The larger x is, the more likely is the event $\{X \leq x\}$. The capital subscript X on $F_X(x)$ denotes that this function F_X is the cdf of the random variable X.

c) The domain of F_X is the range of the random variable X. That is, the random variable X is a mapping from the sample space S to the range R_x. The function F_x is a mapping from R_x to the range of F_x, namely, $[0, 1]$.

Properties of $F_X(x)$

1. $0 \leq F_X(x) \leq 1$ [since $F_X(x)$ is a probability].
2. $F_X(\infty) = 1$, $F_X(-\infty) = 0$.
3. $F_X(x_1) \leq F_X(x_2)$ if $x_1 < x_2$.

These properties follow from the properties of probability. It will be instructive to prove property 3.

Let $x_1 < x_2$. We can write

$$P\{X \leq x_2\} = P\{X \leq x_1\} + P\{x_1 < X \leq x_2\}. \tag{6.4}$$

Rearranging terms, we obtain

$$P\{x_1 < X \leq x_2\} = P\{X \leq x_2\} - P\{X \leq x_1\}$$
$$= F_X(x_2) - F_X(x_1), \tag{6.5}$$

but $P\{x_1 < X \leq x_2\} \geq 0$ since it is a probability. From this and Eq. (6.5) we obtain property 3.

Definition 3. *Probability density function (pdf).* $f_X(x)$ is the pdf of the random variable X. It is defined by

$$f_X(x) = \frac{d}{dx} F_X(x). \tag{6.6}$$

The student who is unfamiliar with the terms cdf and pdf should be warned that there is danger of confusing them. The term *density* has the same meaning here as in Chapters 2 and 3. Recall that we must integrate the power spectral density function $G(f)$ of Chapter 2 to obtain power, and must integrate the energy spectral density function $W(f)$ of Chapter 3 to obtain energy. Now we must integrate the probability density function to obtain probability. For example, the probability that the random variable X has value between x_1 and x_2 is given by

$$P\{x_1 < X < x_2\} = \int_{x_1}^{x_2} f_X(x)\, dx. \tag{6.7}$$

110   RANDOM VARIABLES   6.2

Properties of $f_X(x)$

1. $f_X(x) \geq 0$ for all x.

2. $\int_{-\infty}^{+\infty} f_X(x)\, dx = 1$.

3. $F_X(x) = \int_{-\infty}^{x} f_X(\lambda)\, d\lambda$.

These properties follow from the properties of the cdf.

Relationship to the Sample Space

The domain of the functions F_X and f_X includes the values of the random variable X, that is, includes the range of X. The random variable X is in turn a function whose domain is the sample space. Figure 6.3 illustrates this composite function. Suppose an experiment is performed and the outcome is ζ_i. The random variable X assigns a number of the range, say x_i, to ζ_i, and the function F_X assigns another number, say a_i, to this number x_i. Thus the cdf F_X is a function of the experimental outcome ζ_i. Likewise, the pdf f_X is a function of ζ_i.

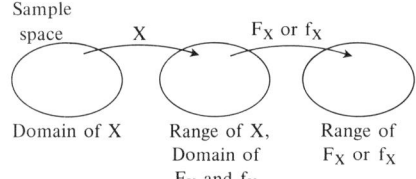

Domain of X   Range of X, Domain of F_X and f_X Range of F_X or f_X

Fig. 6.3

Examples

4. The cdf and pdf of the random variable X of Example 1 are shown in Fig. 6.4. Since F_X is discontinuous at $x = -1$ and $x = 1$, the density function f_X is, strictly speaking, not defined for these values of x. However, we will follow the engineering practice of representing f_X by Dirac delta functions of area equal to the jump (discontinuity) at these values of x. Since we must integrate the function f_X to obtain probability, this practice should not get us into trouble.

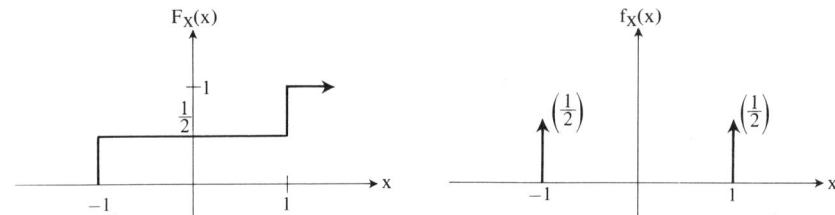

Fig. 6.4

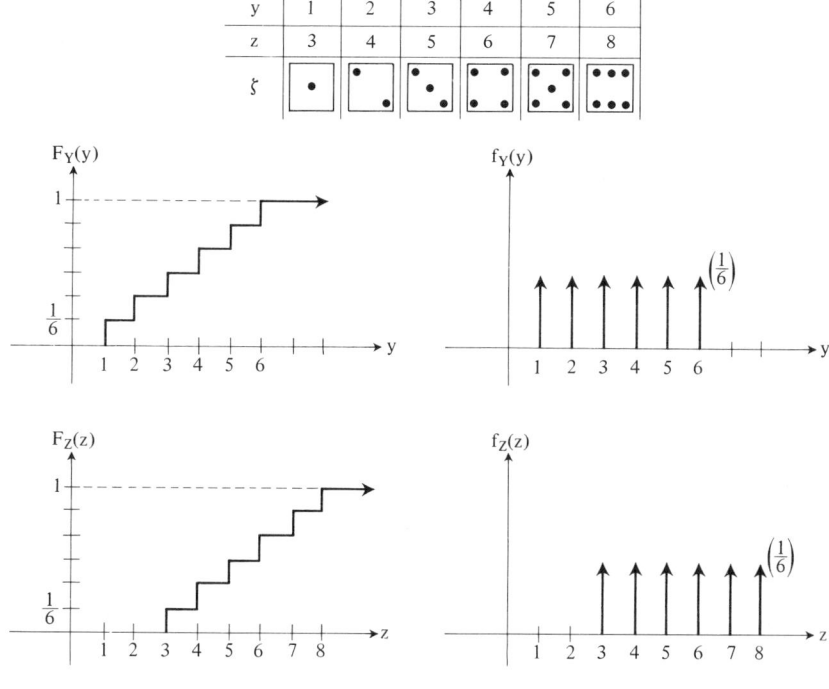

Fig. 6.5

5. Our experiment consists of throwing a single die. There are six possible experimental outcomes. The cdf and pdf of the random variables Y and Z of Example 2 are plotted in Fig. 6.5.

The cdf is a probability. In Examples 4 and 5 the cdf can be determined from the experiment. That is, the four experimental outcomes of Eq. (6.1) are equiprobable, and we can determine $P(\zeta_i)$ for all i. Also the six experimental outcomes of Example 2 are equiprobable. Thus the cdf and pdf are easily determined. However, situations commonly arise, in which the probabilities of the events in the underlying physical experiment are not known. The voltage $v(t_1)$ of the audio amplifier in Example 3 is an example of such a situation.

We can reason, however, that the voltage $v(t_1)$ is most likely to be near zero, and that large values of voltage (both positive and negative) are less likely to occur. Therefore, the cdf and pdf curves should resemble those of Fig. 6.6, and experimental evidence shows this to be the case. In the figure we have used the symbol v in place of $v(t_1)$. The student should keep in mind that the plots given in the figure are for values of voltage v measured at just one time t_1, and that they are not functions of time.

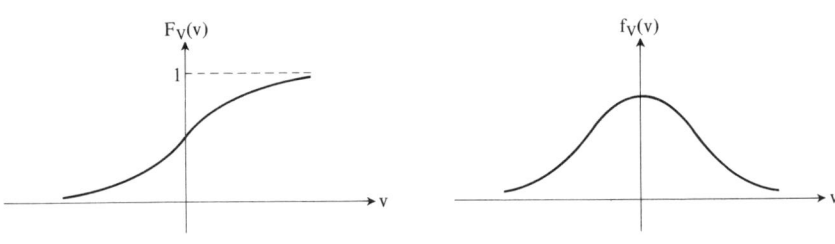

Fig. 6.6

Discrete and Continuous Random Variables

A random variable is termed discrete or continuous depending on whether the cdf and pdf are discrete or continuous. A discrete random variable has steps (discontinuities) in the cdf and impulses in the pdf. A continuous random variable has a continuous cdf and therefore no impulses in the pdf.

Discrete and continuous sample spaces were discussed in Chapter 5.* Since the sample space is the domain of the random variable, the discrete or continuous nature of the random variable must inherently depend on the nature of the sample space. If the sample space is discrete, the random variable must be discrete. If the sample space is continuous, the random variable may be either discrete or continuous.

To illustrate this statement, let us consider the experiment of spinning a pointer on a circle marked off from 0 to 1. (See Fig. 5.3.) We can define both a discrete random variable X and a continuous random variable Y on the sample space $\{x:0 < x < 1\}$. Let X be defined as follows: Assign the number 1 to the set $\{x:0 < x \leq \frac{1}{2}\}$. Assign the number -1 to the set $\{x:\frac{1}{2} < x < 1\}$. This is a discrete random variable. An obvious example of a continuous random variable Y is the identity function $y = x$.

6.3 RELATIONSHIP TO ELECTRICAL ENGINEERING

In this section we attempt to illustrate the relationship of the cdf and pdf to signal waveforms by a series of examples. Suppose the signal waveforms are in the form of voltage. These voltage waveforms vary with time, but the cdf and pdf cannot account for this time change. We must stop time at some particular (but arbitrary) value, say t_1, and measure the voltage $v(t_1)$. Since this time t_1 is arbitrary, the value $v(t_1)$ must be a random quantity.

The random variable V is defined as in Example 3. That is, V is a function with a domain of events "potential difference" and with a range of values

---

* The sample space is discrete if there is a finite or countable number of possible experimental outcomes. The sample space is continuous if there is an uncountably infinite number of possible outcomes.

6.3 RELATIONSHIP TO ELECTRICAL ENGINEERING

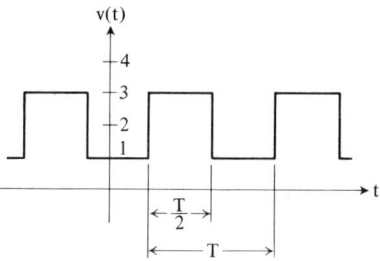

Fig. 6.7

$v(t_1)$. The pdf and cdf will depend on the manner in which $v(t)$ varies with time, although the pdf and cdf will not themselves be time functions. Several illustrations are given below.

1. *Discrete distribution.* Suppose $v(t)$ is the periodic waveform shown in Fig. 6.7. Since the value of $v(t)$ is 3 half of the time and 1 the other half, the voltmeter reading $v(t_1)$ must be either 1 or 3, and these values are equally likely.

Now we wish to find the value of the function $F_V(v)$ for different values of v. For $v \geq 3$ the event $\{V \leq v\}$ is certain (the voltage must be 1 or 3 and either value is less than or equal to v). Therefore $P\{V \leq v\} = 1$ for $v \geq 3$.

For values of v in the range $1 \leq v < 3$, the probability $P\{V \leq v\} = \frac{1}{2}$. The random variable V can assume only the values 1 or 3. If $V = 3$, then it is not true that $V \leq v$. If $V = 1$, it is true.

Finally, for $v < 1$ it can never happen that $V \leq v$. Therefore $P\{V \leq v\} = 0$ for $v < 1$.

The function $F_V(v)$ is, for all values of v, given by

$$F_V(v) = P\{V \leq v\} = \begin{cases} 1, & 3 \leq v, \\ \frac{1}{2}, & 1 \leq v < 3, \\ 0, & v < 1. \end{cases} \quad (6.8)$$

The cdf and pdf are plotted in Fig. 6.8. The pdf $f_V(v)$ is found by differentiating $F_V(v)$, as in Eq. (6.6). Since this random variable V is discrete we

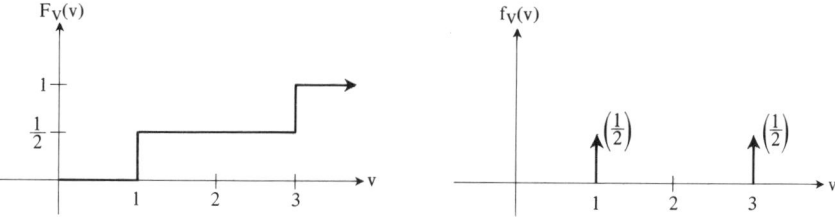

Fig. 6.8

114   RANDOM VARIABLES   6.3

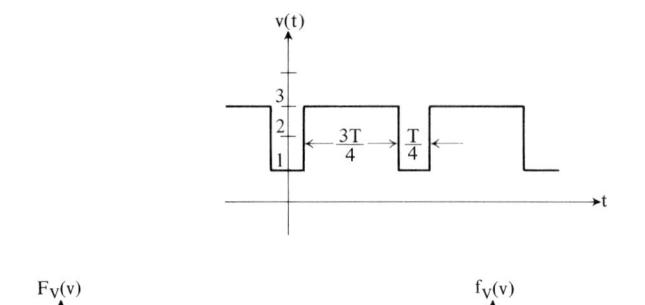

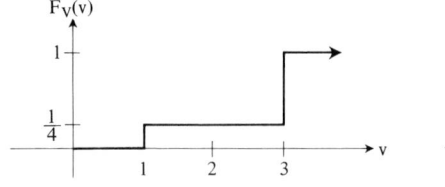
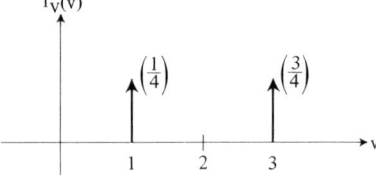

Fig. 6.9

could have found $f_V(v)$ directly, for the value $P\{V = 1\} = \frac{1}{2}$ represents the area under the curve $f_V(v)$ in the neighborhood of $v = 1$. The value $P\{V = 3\} = \frac{1}{2}$ represents the area under $f_V(v)$ in the neighborhood of $v = 3$.

2. Figure 6.9 illustrates another voltage waveform with the corresponding cdf and pdf. Again, the random variable is discrete, with possible values of 1 and 3. The probabilities of these two values are no longer equal, however. From the diagram of the waveform $v(t)$ it is evident that $P\{V = 3\} = \frac{3}{4}$ and $P\{V = 1\} = \frac{1}{4}$. The associated cdf and pdf should be compared to those of Fig. 6.8 to illustrate the effect of these different probabilities. The cdf is given by

$$F_V(v) = \begin{cases} 1, & 3 \le v, \\ \frac{1}{4}, & 1 \le v < 3, \\ 0, & v < 1. \end{cases} \tag{6.9}$$

3. *Continuous distribution.* We now encounter a continuous random variable. For this type of random variable there are no jumps (discontinuities) in the cdf and no delta functions in the pdf.

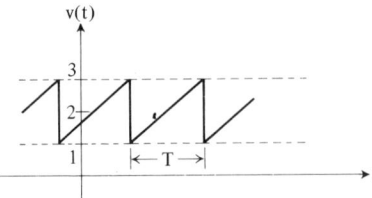

Fig. 6.10

Consider the waveform $v(t)$ shown in Fig. 6.10. At any time t_1 the value of the voltage $v(t_1)$ will be in the range $1 \le v(t_1) \le 3$. Therefore we can state immediately that

$$F_V(v) = \begin{cases} 1, & 3 < v, \\ 0, & v < 1. \end{cases} \qquad (6.10)$$

Our only problem is to compute the function F_V for values of v in the range [1, 3]. But we can see that the function is linear in this range. Suppose $v = 2$. Then $\{V \le v\}$ will occur one-half of the time. If $v = 2.5$, then $\{V \le v\}$ will occur three-fourths of the time, and if $v = 1.5$, then $\{V \le v\}$ will occur one-fourth of the time. Thus

$$F_V(v) = \tfrac{1}{2}(v - 1), \qquad 1 \le v < 3. \qquad (6.11)$$

Figure 6.11 illustrates this cdf and the corresponding pdf. This type of distribution is called *uniform*.

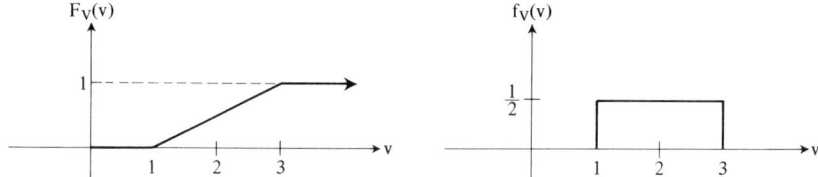

Fig. 6.11

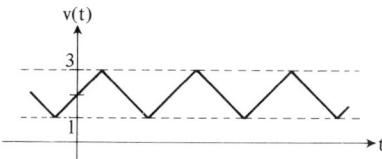

Fig. 6.12

4. The voltage waveform shown in Fig. 6.12 has the cdf and pdf shown in Fig. 6.11.

5. *Mixed distribution.* The waveform shown in Fig. 6.13(a) has mixed distribution (both continuous and discrete). There are discontinuities in the cdf (Fig. 6.13b) and impulses in the pdf (Fig. 6.13c). The cdf is given by

$$F_V(v) = \begin{cases} 1, & 2 \le v, \\ \tfrac{1}{3}(v + 1), & 0 \le v < 2, \\ 0, & v < 0. \end{cases} \qquad (6.12)$$

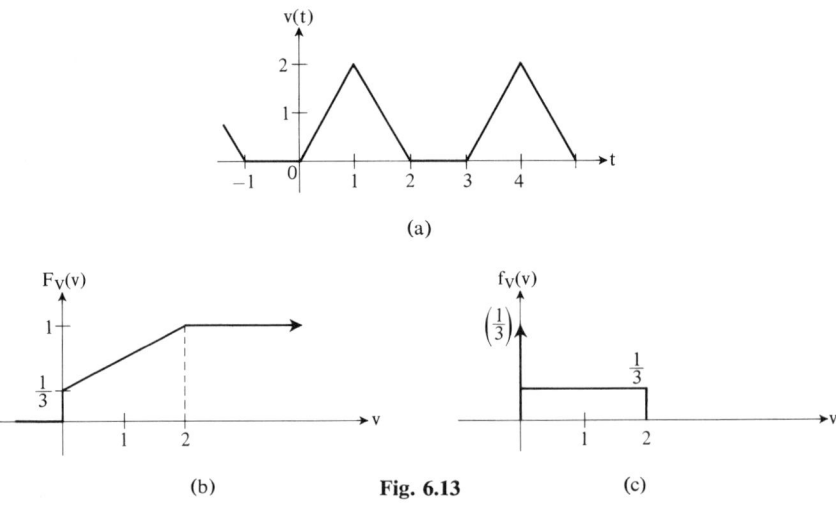

Fig. 6.13

These five illustrations have been for waveforms $v(t)$ that are linear, that is, the graphs of $v(t)$ are straight lines. Where the slope of the line is zero we have a discrete random variable. Otherwise the random variable is uniform. To illustrate that other types of distributions can arise, we shall study a common waveform.

6. *Sinusoidal distribution.* If the sine wave shown in Fig. 6.14(a) is sampled at an arbitrary time t_1, then the cdf and pdf will be as shown in Fig. 6.14(b) and 6.14(c). This distribution will be derived in Chapter 7.

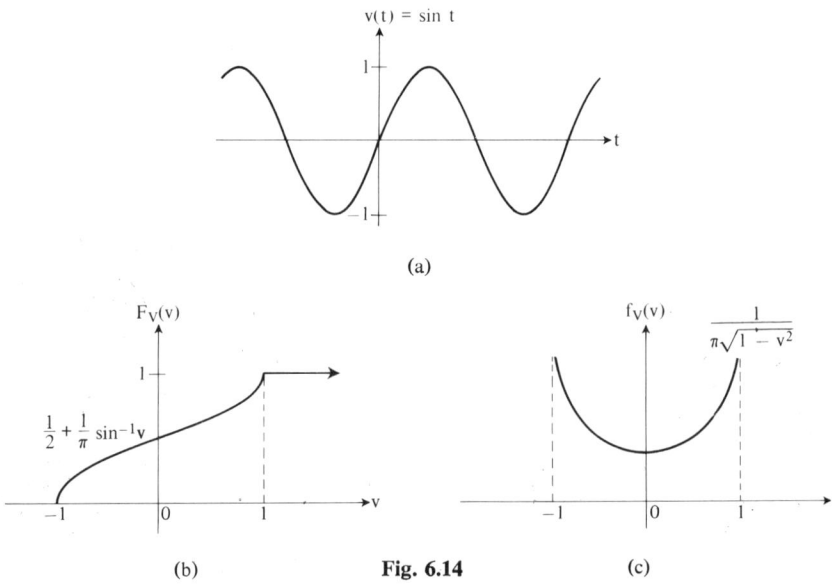

Fig. 6.14

6.4 STATISTICAL AVERAGES

Consider a random voltage waveform such as the oscillogram of a voice recording. Figure 6.15 shows such a waveform. What can we say about this waveform to describe it mathematically? Not much! We certainly cannot write a formula that will describe the waveform. About all we can do is measure the dc-level (average value) and average power, and these two numbers hardly describe the waveform.

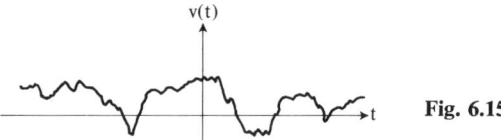

Fig. 6.15

Fortunately, there are a few other numbers that can be derived from the waveform that are of some use. This section is devoted to describing these numbers, and the rest of this text is devoted to describing their use.

Definition 4. *Expected value.* The expected value of a random variable X is $E(X)$. It is given by

$$E(X) = \int_{-\infty}^{+\infty} x f_X(x)\, dx. \quad (6.13)$$

Other names for $E(X)$ are average value, mean value, and statistical average.

This definition may readily be extended to any function of X, say $g(X)$, by

$$E[g(X)] = \int_{-\infty}^{+\infty} g(x) f_X(x)\, dx. \quad (6.14)$$

For example, if $g(X) = X^2$, then the mean-square value of the random variable X is

$$E(X^2) = \int_{-\infty}^{+\infty} x^2 f_X(x)\, dx. \quad (6.15)$$

Note that the average value of $g(X)$ is a functional. The domain consists of functions $g(X)$ and the range is a set of numbers, the expected values of $g(X)$. For different functions we get different numbers (generally).

Moments

The average value of X is the first moment m_1. The mean-square value $E(X^2)$ is the second moment m_2. In general, the nth moment is given by

$$m_n = E(X^n) = \int_{-\infty}^{+\infty} x^n f_X(x)\, dx. \quad (6.16)$$

118 RANDOM VARIABLES 6.4

Another important term is *central moment*. If we first subtract m_1 (the mean) from X and then find the average value of $g(X) = (X - m_1)^n$, we get the nth central moment. The second central moment is so important that it rates a special name, *variance*, and a special label, σ^2. The variance of X is given by

$$\sigma^2 = E[(X - m_1)^2] = \int_{-\infty}^{+\infty} (x - m_1)^2 f_X(x) \, dx. \tag{6.17}$$

The square root of the variance is called the *standard deviation*, σ.

Relationship of Moments to Electrical Engineering

For random variables related to voltage waveforms (such as those in Section 6.3), the first moment m_1 is the dc-value. Additional relationships among parameters for waveforms are

$$\begin{aligned} m_1 &= \text{dc-value,} \\ m_1^2 &= \text{dc-power,} \\ \sigma^2 &= \text{ac-power,} \\ m_2 &= \text{total power} = \sigma^2 + m_1^2. \end{aligned} \tag{6.18}$$

This last relationship can be demonstrated as follows. By definition,

$$\sigma^2 = E[(V - m_1)^2] = E(V^2 - 2m_1 V + m_1^2).$$

Then

$$\begin{aligned} \sigma^2 &= E(V^2) - 2m_1 E(V) + m_1^2 = E(V^2) - m_1^2 \\ &= m_2 - m_1^2. \end{aligned} \tag{6.19}$$

The relationships given in Eqs. (6.18) simply state that if the waveform is decomposed by the subtraction of the dc-level, then m_1^2 is the power in a dc-signal of that amplitude; the variance σ^2 is the power in the signal after the dc-level is removed; and the second moment m_2 is the power in the original waveform. We assume, of course, that we are dealing with a power signal.

For the waveforms discussed in Section 6.3 we can compute the dc-power, ac-power, and total power by old familiar techniques (Chapter 2). But we can also use this new (statistical) technique. In many cases we cannot describe the waveform by a simple formula, but if the cdf or pdf is known, we can still compute the power.

Below are some sample calculations for the waveforms given in Section 6.3. These illustrations are numbered to correspond with those of Section 6.3. The student should verify that the dc-level, dc-power, ac-power, and total power obtained are correct by making these computations using the techniques of Chapter 2.

1. The pdf is given by

$$f_V(v) = \tfrac{1}{2}\delta(v - 1) + \tfrac{1}{2}\delta(v - 3).$$

6.4 STATISTICAL AVERAGES

The average value is

$$m_1 = \int_{-\infty}^{+\infty} v f_V(v)\, dv = \int_{-\infty}^{+\infty} v[\tfrac{1}{2}\delta(v-1) + \tfrac{1}{2}\delta(v-3)]\, dv$$
$$= \tfrac{1}{2}(1) + \tfrac{1}{2}(3) = 2.$$

The variance (ac-power) is

$$\sigma^2 = \int_{-\infty}^{+\infty} (v - m_1)^2 f_V(v)\, dv = \int_{-\infty}^{+\infty} (v-2)^2[\tfrac{1}{2}\delta(v-1) + \tfrac{1}{2}\delta(v-3)]\, dv$$
$$= \tfrac{1}{2}(1) + \tfrac{1}{2}(1) = 1.$$

And the mean-square value (total power) is

$$m_2 = \int_{-\infty}^{+\infty} v^2 f_V(v)\, dv = \int_{-\infty}^{+\infty} v^2[\tfrac{1}{2}\delta(v-1) + \tfrac{1}{2}\delta(v-3)\, dv]$$
$$= \tfrac{1}{2}(1) + \tfrac{1}{2}(9) = 5.$$

Note that $m_2 = \sigma^2 + m_1^2$.

2. The pdf is given by

$$f_V(v) = \tfrac{1}{4}\delta(v-1) + \tfrac{3}{4}\delta(v-3).$$

The moments m_1, σ^2, and m_2 are computed as above and these computations will not be repeated here. The values obtained are

$$m_1 = 2.5, \qquad \sigma^2 = \tfrac{3}{4}, \qquad m_2 = 7 = m_1^2 + \sigma^2.$$

3. The pdf is given by

$$f_V(v) = \begin{cases} \tfrac{1}{2} & \text{if } 1 < v < 3, \\ 0 & \text{otherwise.} \end{cases}$$

The average value is

$$m_1 = \int_{-\infty}^{+\infty} v f_V(v)\, dv = \int_1^3 v(\tfrac{1}{2})\, dv = 2.$$

The variance (ac-power) is

$$\sigma^2 = \int_{-\infty}^{+\infty} (v - m_1)^2 f_V(v)\, dv = \int_1^3 (v-2)^2(\tfrac{1}{2})\, dv = \tfrac{1}{3}.$$

And the second moment (total power) is

$$m_2 = \int_{-\infty}^{+\infty} v^2 f_V(v)\, dv = \int_1^3 v^2(\tfrac{1}{2})\, dv = \tfrac{13}{3}.$$

Again a check shows that $m_2 = \sigma^2 + m_1^2$.

The computation of these parameters for the remaining waveforms of Section 6.3 is left to the reader as a worthwhile exercise.

Note. It is convenient to use a single notation for the expected value of both discrete and continuous random variables. We are referring to the integrals of Eqs. (6.13) through (6.16). However, it is less cumbersome to work with the sum of numbers instead of the integral of delta functions, so for discrete random variables we introduce the following convention.

Refer to illustration 1 on page 118. Replace the two delta functions in the pdf by the probabilities

$$P\{V = 1\} = \tfrac{1}{2}, \qquad P\{V = 3\} = \tfrac{1}{2}.$$

The average value is computed by the sum

$$m_1 = \sum_v v P\{V = v\} = 1 \cdot \tfrac{1}{2} + 3 \cdot \tfrac{1}{2} = 2$$

rather than by the integral of the delta functions as in the illustration. The variance and mean-square value can be computed in a similar manner. For example, the variance is given by

$$\sigma^2 = \sum_v (v - m_1)^2 P\{V = v\}$$
$$= (1 - 2)^2 \cdot \tfrac{1}{2} + (3 - 2)^2 \cdot \tfrac{1}{2} = 1.$$

In the following we will compute averages for discrete random variables in this manner without comment.

6.5 SOME IMPORTANT DISTRIBUTIONS

In Section 6.3 we saw that the probability distribution of the amplitude of a waveform was determined by the manner in which the waveform varied with time. This is true for random waveforms that cannot be described by an equation, just as it was true for the waveforms in the above examples. The most common probability distribution for random waveforms is the gaussian distribution. We now discuss this distribution, along with three other important distributions: the binary, binomial, and uniform distributions.

Binary Distribution

This distribution is also called Bernoulli distribution. The random variable in Example 4 has binary distribution. There are two numbers in the range of a binary random variable. In Example 4 these numbers are -1 and $+1$.

In general, the random variable partitions the sample space into two distinct subsets, say A and B. All elements of one subset, A, are mapped into one number, say a, and all elements of the other (complementary) set, B, are mapped into another number, say b. That is,

$$X(\zeta_i) = \begin{cases} a & \text{if } \zeta_i \in A, \\ b & \text{if } \zeta_i \in B. \end{cases}$$

(6.20)

(6.21)

6.5 SOME IMPORTANT DISTRIBUTIONS

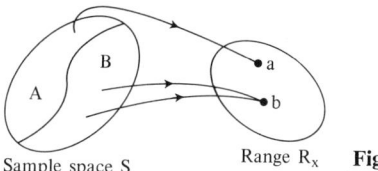

Fig. 6.16

Figure 6.16 is a graphic representation of a binary random variable. Each element in the sample space S is mapped into one of two numbers, a or b.

An important example of this random variable is the "success" or "failure" random variable. An experiment is performed. If it results in "success," the random variable X assigns the number 1, if it results in "failure," $X = 0$. For example, we can toss a coin and assign the number 1 to heads and the number 0 to tails. Or, on a production line, we can test a part to see if it meets specifications. If so, $X = 1$; if not, $X = 0$. Similarly, we can transmit a binary (on-off) message sequence. If "on," $X = 1$; if "off," $X = 0$.

The probability of "success" is usually denoted by p, and the probability of "failure" by $1 - p = q$. Thus

$$P\{X = 1\} = p, \tag{6.22}$$

$$P\{X = 0\} = q. \tag{6.23}$$

The mean and variance of this random variable are p and pq, respectively. Figure 6.17 illustrate its cdf and pdf.

Binomial Distribution

Suppose we perform the binary experiment n times with outcomes $X_1, X_2, \ldots, X_n$, where X_i is either 0 or 1. Then we have n random variables $X_1, X_2, \ldots, X_n$, where X_i is associated with the ith trial. Now we define the random variable Y as the sum of the X_i's:

$$Y = \sum_{i=1}^{n} X_i. \tag{6.24}$$

Since the range of each X_i is 0 or 1, the range of Y is the set of integers $\{0, 1, \ldots, n\}$. That is, Y can assume any one of the numbers $\{0, 1, \ldots, n\}$,

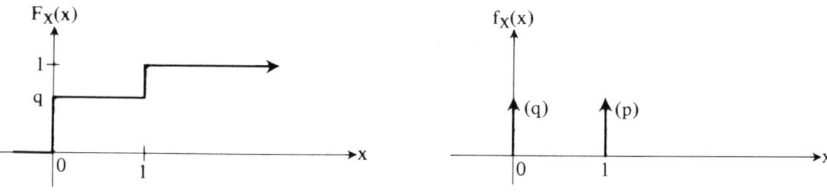

Fig. 6.17 The cdf and pdf for the "success" or "failure" binary random variable.

and the probability that Y will be equal to a particular number is determined by the two probabilities $P\{X = 1\}$ and $P\{X = 0\}$ and by the number n. Given that $P\{X_i = 1\} = p$ and $P\{X_i = 0\} = q$, we wish to find the cdf and pdf of the random variable Y. To do this we will need Pascal's triangle.

```
                    1     1                    1
                 1     2     1                 2
              1     3     3     1              3
           1     4     6     4     1           4
        1     5    10    10     5     1        5
     1     6    15    20    15     6     1     6
```

Fig. 6.18 Pascal's triangle.

Pascal's triangle contains the coefficients in the binomial expansion $(p + q)^n$. Figure 6.18 shows Pascal's triangle for values of n from 1 to 6. For example, the binomial expansion of $(p + q)^4$ is

$$(p + q)^4 = p^4 + 4p^3q + 6p^2q^2 + 4pq^3 + q^4. \tag{6.25}$$

Note that the coefficients of the terms in Eq. (6.25) are given in the fourth row of the triangle.

This triangle is constructed in the following way. Each row begins and ends with the number 1. The numbers in between are determined by adding the two numbers in the row just above the desired number. For example, the numbers in the fourth row are $4 = 1 + 3$, $6 = 3 + 3$, and $4 = 3 + 1$. In this manner the seventh row could easily be added to Fig. 6.18. It would be 1, 7, 21, 35, 35, 21, 7, 1.

Another way to construct Pascal's triangle is to use the *binomial coefficient* $\binom{n}{k}$, which is the symbol for the following operation:

$$\binom{n}{k} = \frac{n!}{k!(n-k)!}. \tag{6.26}$$

The first four rows of Pascal's triangle are shown in Fig. 6.19 in terms of this new notation. As an example of the use of the binomial coefficient, note that the center coefficient in the fourth row is given by

$$\binom{4}{2} = \frac{4!}{2!\,2!} = \frac{4 \cdot 3 \cdot 2 \cdot 1}{(2 \cdot 1)(2 \cdot 1)} = 6. \tag{6.27}$$

Pascal's triangle and the binomial coefficients are useful because they enable us to find the number of ways to arrange n things taken k at a time. For example, suppose we toss a coin four times and want to know the number of different ways we can obtain two heads. The answer is given by Eq. (6.27):

there are six ways to obtain two heads if a coin is tossed four times (list them). Thus $\binom{n}{k}$ is the number of ways we can obtain k successes in n trials of the experiment.

$$
\begin{array}{ccccccc}
 & & & \binom{1}{0} & \binom{1}{1} & & & & 1 \\
 & & \binom{2}{0} & & \binom{2}{1} & & \binom{2}{2} & & 2 \\
 & \binom{3}{0} & & \binom{3}{1} & & \binom{3}{2} & & \binom{3}{3} & 3 \\
\binom{4}{0} & & \binom{4}{1} & & \binom{4}{2} & & \binom{4}{3} & & \binom{4}{4} & 4
\end{array}
$$

Fig. 6.19 Pascal's triangle.

In order for the random variable Y of Eq. (6.24) to equal K, where K is one of the numbers $\{0, 1, \ldots, n\}$, the experiment must result in K successes and $(n - K)$ failures. If the trials are independent, the probability of K successes and $(n - K)$ failures in any given order is the product of K p's and $(n - K)$ q's. That is, this probability is equal to $p^K q^{(n-K)}$. Since there are $\binom{n}{K}$ different ways to obtain K successes and $(n - K)$ failures, the distribution of Y is given by

$$P\{Y = K\} = \binom{n}{K} p^K q^{n-K}. \tag{6.28}$$

Example 6. If a single die is thrown four times, what is the probability that a 1 will turn up twice? Here $p = \frac{1}{6}$ and $q = \frac{5}{6}$. (Success is rolling a 1.) Therefore,

$$P\{Y = 2\} = \binom{4}{2} p^2 q^2 = 6 \left(\frac{1}{6}\right)^2 \left(\frac{5}{6}\right)^2 = \frac{25}{216}. \tag{6.29}$$

The cdf and pdf of the random variable Y of this example are plotted in Fig. 6.20. Figure 6.21 shows a plot of the binomial distribution for the

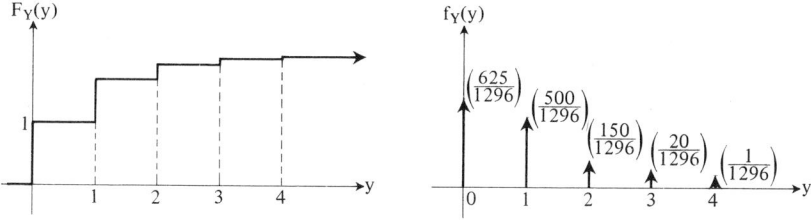

Fig. 6.20 The binomial distribution for $n = 4$, $p = \frac{1}{6}$.

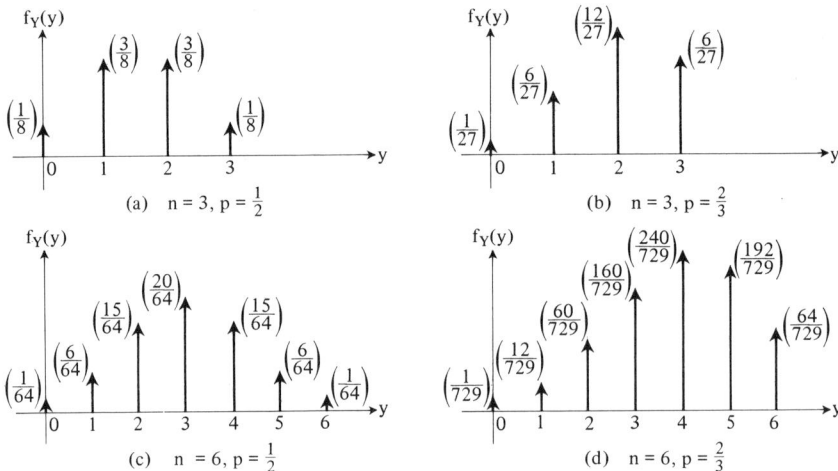

Fig. 6.21 The binomial pdf for various values of n and p.

indicated values of n and p. The particular distributions illustrated in Fig. 6.21 could arise from many different experiments. One possibility for each distribution is as follows: If a fair coin is tossed three times with success defined as "head," the pdf of Fig. 6.21(a) will result. In Fig. 6.21(b) a die could be rolled three times where "success" is rolling either a 1, 2, 3, or 4. If the coin is tossed six times, the pdf in Fig. 6.21(c) will occur, so the only difference between the experiments for (a) and (c) is the number of trials. Likewise, the only difference between the experiments for (b) and (d) is the number of trials.

Note. The mean and variance of the binomial random variable are np and npq, respectively.

Uniform Distribution

The binary and binomial distributions are both discrete; that is, the random variable can assume only discrete values. We now introduce two continuous distributions: the uniform and gaussian distributions.

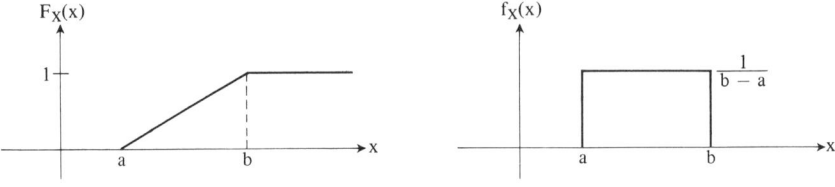

Fig. 6.22

6.5 SOME IMPORTANT DISTRIBUTIONS

A random variable X is *uniformly distributed* if the pdf is given by

$$f_X(x) = \begin{cases} \dfrac{1}{b-a} & \text{if } a < x < b, \\ 0 & \text{otherwise.} \end{cases} \qquad (6.30)$$

We then say that X is uniformly distributed in the interval (a, b). Figure 6.22 illustrates the cdf and pdf for a random variable that is uniformly distributed between values a and b. Since the total area under $f_X(x)$ must be one, the height of the uniform pdf is determined by its width.

The mean and variance of the uniformly distributed random variable are $\frac{1}{2}(b + a)$ and $\frac{1}{12}(b - a)^2$, respectively. Note that the numbers for the mean and variance correspond to those in the illustrations of the previous section.

Gaussian Distribution

In our mathematical models for random phenomena the *gaussian* (also called *normal*) *distribution* occurs most often. This is a result of two factors: the central limit theorem, and the ease with which the gaussian function can be manipulated.

This ease of manipulation is an important factor. In many problems that we investigate, we can find a general solution although we may have no knowledge of the distribution of the random variable involved. However, specific solutions depend on the distribution of the random variable, and often the gaussian random variable is the only one that we can manipulate to find a solution.

The central limit theorem provides us with a theoretical justification for using the gaussian distribution. It states, in effect, that the gaussian random variable is an appropriate model for many of the random phenomena found in nature. We will discuss the central limit theorem in Chapter 8.

A random variable X has gaussian distribution if the pdf is given by

$$f_X(x) = \frac{1}{\sqrt{2\pi\sigma_X^2}} e^{-(x-m_X)^2/2\sigma_X^2}. \qquad (6.31)$$

The factors m_X and σ_X^2 are the mean and variance of the random variable. The associated cdf is then given by

$$F_X(x) = \int_{-\infty}^{x} \frac{1}{\sqrt{2\pi\sigma_X^2}} e^{-(\alpha-m_X)^2/2\sigma_X^2} \, d\alpha. \qquad (6.32)$$

These functions are illustrated in Fig. 6.23.

Notes. a) The mean m_X locates the gaussian curve along the x-axis, and the variance σ_X^2 is related to the width of the pdf.

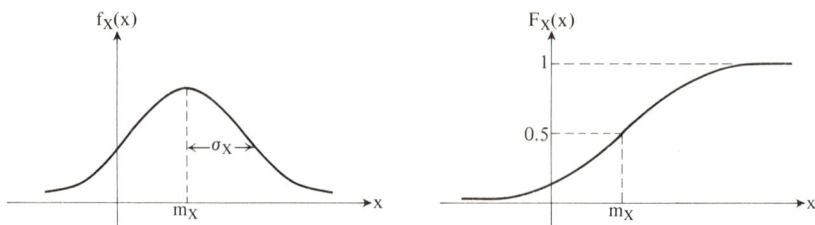

Fig. 6.23 The pdf and cdf for the gaussian random variable.

b) Note that the gaussian random variable is completely characterized by its first two moments, m_X and σ_X^2. That is, if the mean and variance of a gaussian random variable are known, then the distribution is known.

c) The integral in Eq. (6.32) cannot be evaluated directly. Most statistics and probability texts provide a table of values of the cdf $F_X(x)$, and a digital computer can be used to evaluate this integral. (See the Table of Normal Curve Areas at the back of the book.)

Example 7. Suppose that a random waveform $v(t)$ can be sampled at times $\{t_i\}$ and that the values $\{v(t_i)\}$ have gaussian distribution with zero mean and unit variance. This is written $f_X(x) \sim N(0, 1)$, which means that the random variable X is normal with zero mean and unit variance.

a) Find the probability that a particular value $v(t_i)$ will exceed its rms value.
b) Find the probability that $\{|v(t_i)| < \frac{1}{2}\}$.

From Eq. (6.18) the rms value of $v(t)$ is just the standard deviation σ (since $m_1 = 0$). Thus part (a) asks for $P\{v(t_i) > 1\}$. This is the shaded area in Fig. 6.24(a), and is given by $1 - F_X(1)$. From the Table of Normal Curve Areas, $P\{v(t_i) > 1\} = 1 - 0.8413 = 0.1587$.

Part (b) asks for the shaded area of Fig. 6.24(b). The area under the curve from 0 to $\frac{1}{2}$ is 0.1915 (see the Table of Normal Curve Areas). Thus the area under f_V from $-\frac{1}{2}$ to $\frac{1}{2}$ is $2(0.1915) = 0.3830$, and this is $P\{|v(t_i)| < \frac{1}{2}\}$.

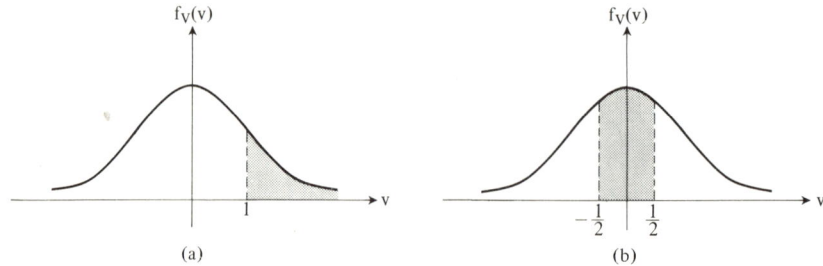

Fig. 6.24

6.6 CHARACTERISTIC FUNCTIONS

The *characteristic function* ϕ_X of the random variable X and the pdf f_X form a Fourier transform pair:

$$\phi_X(v) \leftrightarrow f_X(x). \tag{6.33}$$

The characteristic function has many uses, and we will use it in discussing the central limit theorem, in evaluating moments, and in special cases of transformation of random variables.

All the properties of Fourier transforms discussed in Chapter 3 can be applied to characteristic functions. The Fourier transform pairs in terms of ϕ_X and f_X are

$$\phi_X(v) = \int_{-\infty}^{+\infty} f_X(x) e^{jvx} \, dx, \tag{6.34}$$

$$f_X(x) = \frac{1}{2\pi} \int_{-\infty}^{+\infty} \phi_X(v) e^{-jvx} \, dv. \tag{6.35}$$

Notes. a) Equations (6.34) and (6.35) might look more familiar if you identify x and v with angular frequency and time, respectively. That is, for the correspondence

$$x = \omega, \quad v = t,$$

Eqs. (6.34) and (6.35) are almost identical to Eqs. (3.1) and (3.3). The 2π-factor is associated with the wrong integral; otherwise they are the same.

b) An alternative definition for the characteristic function in terms of expectation is

$$\phi_X(v) = E(e^{jvx}) = \int_{-\infty}^{+\infty} e^{jvx} f_X(x) \, dx. \tag{6.36}$$

This is one reason that the 2π-factor is associated with the wrong integral. Another reason is for ease and simplicity in using ϕ_X to calculate moments (see below).

c) As in the pdf, the subscript X in ϕ_X denotes that this is the characteristic function for the random variable X.

Examples

8. Let X be a binary random variable with the pdf shown in Fig. 6.25(a). Then $\phi_X(v)$ is a cosine waveform, as shown in Fig. 6.25(b), since

$$\phi_X(v) = \int_{-\infty}^{+\infty} [\tfrac{1}{2}\delta(x+1) + \tfrac{1}{2}\delta(x-1)] e^{jvx} \, dx$$

$$= \tfrac{1}{2} e^{-jv} + \tfrac{1}{2} e^{jv} = \cos v. \tag{6.37}$$

9. As a generalization of Example 8, suppose X is a binary random variable with pdf $P\{X = a\} = p$ and $P\{X = b\} = q = 1 - p$; that is,

$$f_X(x) = p\delta(x-a) + q\delta(x-b). \tag{6.38}$$

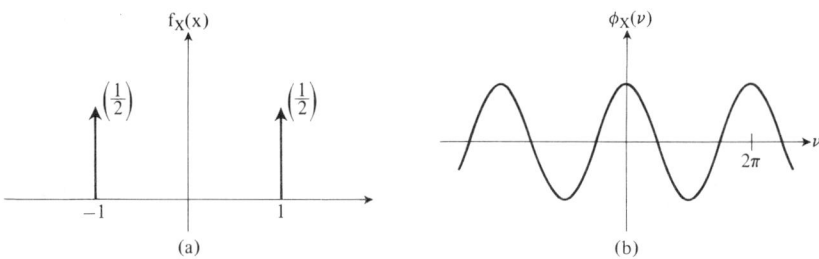

Fig. 6.25

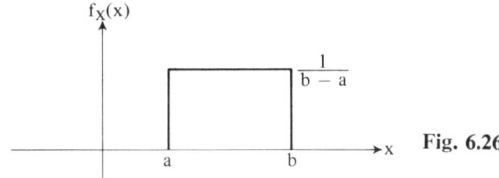

Fig. 6.26

Then
$$\phi_X(v) = pe^{jav} + qe^{jbv}. \quad (6.39)$$

10. If X is uniformly distributed $(-1, 1)$, then
$$\phi_X(v) = \int_{-1}^{+1} \tfrac{1}{2} e^{jvx}\, dx = (\sin v)/v. \quad (6.40)$$

11. As a generalization of Example 10, suppose X is uniformly distributed (a, b). (See Fig. 6.26.) Then
$$\phi_X(v) = \frac{1}{b-a} \int_a^b e^{jvx}\, dx$$
$$= \frac{1}{jv(b-a)} (e^{jvb} - e^{jva}). \quad (6.41)$$

12. One of the distinguishing features of the gaussian curve is that its Fourier transform is also gaussian. For normal distribution with zero mean and unit variance [written $N(0, 1)$], the characteristic function $\phi_X(v)$ is given by
$$\phi_X(v) = e^{-v^2/2} \longleftrightarrow \frac{1}{\sqrt{2\pi}} e^{-x^2/2}. \quad (6.42)$$

First consider the effect of changing the mean. In Fig. 6.27(b) the curve is shifted so that the center is at m_X instead of 0. By the modulation property, Eq. (3.16), the characteristic function is multiplied by e^{jvm_X}. That is,
$$\phi_X(v) = e^{jvm_X} e^{-v^2/2} \longleftrightarrow \frac{1}{\sqrt{2\pi}} e^{-(x-m_X)^2/2}. \quad (6.43)$$

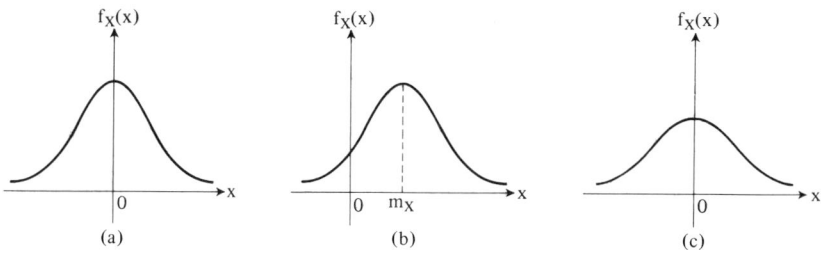

Fig. 6.27 The gaussian curve. (a) Zero mean and unit variance. (b) Mean at m_X and unit variance. (c) Zero mean and variance σ_X^2.

Next consider the effect of changing the variance, as shown in Fig. 6.27(c). By the scaling property, Eq. (3.14), if the variance of X is changed from 1 to σ_X^2, the characteristic function becomes

$$\phi_X(v) = e^{-v^2\sigma_X^2/2} \leftrightarrow \frac{1}{\sqrt{2\pi\sigma_X^2}} e^{-x^2/2\sigma_X^2} \qquad (6.44)$$

If we combine these two effects, the characteristic function for the general normal random variable, $N(m_X, \sigma_X^2)$, becomes

$$\phi_X(v) = e^{jvm_X}e^{-v^2\sigma_X^2/2} \leftrightarrow \frac{1}{\sqrt{2\pi\sigma_X^2}} e^{-(x-m_X)^2/2\sigma_X^2}. \qquad (6.45)$$

Use of the Characteristic Function in Evaluating Moments

Here we use the differentiation property of Fourier transforms, Eq. (3.13). Consider the defining equation for the characteristic function Eq. (6.34). Differentiating with respect to v, we obtain

$$\frac{d}{dv}\phi_X(v) = \int_{-\infty}^{+\infty} jxf_X(x)e^{jvx}\,dx. \qquad (6.46)$$

Evaluating this derivative at $v = 0$, we get

$$\frac{d}{dv}\phi_X(v)\bigg|_{v=0} = j\int_{-\infty}^{+\infty} xf_X(x)\,dx = jm_1. \qquad (6.47)$$

Or, dividing by j, we have

$$m_1 = \frac{1}{j}\frac{d}{dv}\phi_X(0). \qquad (6.48)$$

If this procedure is continued, the nth derivative of $\phi_X(v)$ evaluated at $v = 0$ can be used to obtain the nth moment:

$$m_n = \frac{1}{(j)^n}\frac{d^n}{dv^n}\phi_X(0). \qquad (6.49)$$

Other uses for the characteristic function will be investigated later.

Example 13. In Section 6.5 we stated that the mean and variance of the success or failure binary random variable in Fig. 6.17 were p and pq, respectively. From Eq. (6.39), the characteristic function is given by

$$\phi_X(v) = q + pe^{jv}.$$

Using Eq. (6.49), we get

$$m_1 = \frac{1}{j}(jp) = p, \qquad m_2 = \frac{1}{j^2}(j^2 p) = p,$$

so that $\sigma^2 = m_2 - m_1^2 = p - p^2 = p(1 - p) = pq$.

PROBLEMS

1. Define the random variable X as in the following table.

X	-3	+2	+1	0	-1	-2
ζ	⚀	⚁	⚂	⚃	⚄	⚅

a) Compute and plot $F_X(x)$.
b) Compute and plot $f_X(x)$.

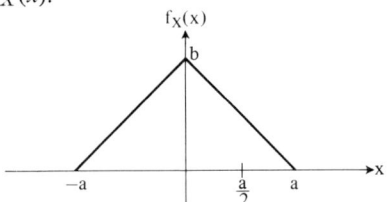

Fig. 6.28

2. Consider the triangular pdf shown in Fig. 6.28.
a) Find b in terms of a.
b) Plot the cdf.
c) What is $P(X > a/2)$?
3. Given that X is a random variable, find the probability that
a) $a < X < b$,   b) $a \le X \le b$,   c) $a < X \le b$
in terms of the distribution function of X.
4. Repeat Problem 3 in terms of the pdf, given that X is a continuous random variable.
5. Suppose you sample the voltage of the periodic waveforms given in Fig. 6.29 at an arbitrary time t_1. Plot the pdf and cdf of each waveform for the voltage $v(t_1)$.
6. Let X be a random variable distributed as in Problem 5(d). Find
a) $P(X = 1)$,   b) $P(X = 0)$,   c) $P(X < 1)$.

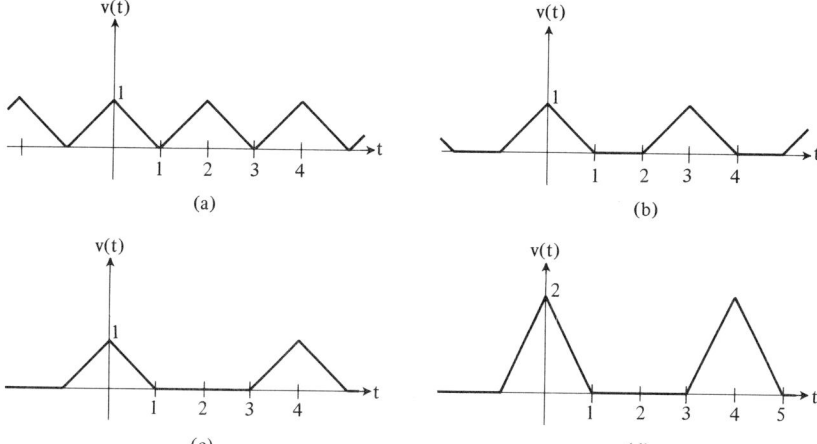

Fig. 6.29

7. The random variable Y has pdf $f_Y(y) = ae^{-b|y|}$ for $-\infty < y < +\infty$.
   a) Determine the required relationship between a and b.
   b) Determine the mean and variance of Y.

8. Given that X is a random variable taking on the values x_1, x_2, and x_3 with probabilities p_1, p_2, and p_3, find the expression for the (a) mean, (b) variance.

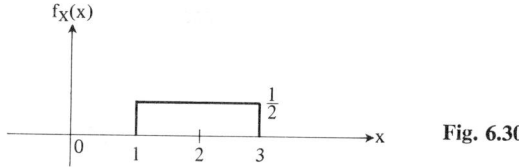

Fig. 6.30

9. Suppose the random variable X is uniformly distributed $(1, 3)$ (see Fig. 6.30).
   a) Find the average value of X.
   b) Find the mean-square value of X.
   c) Find $P(X > 2.5)$.

10. Repeat Problem 9 given that X is normal $N(2, 4)$.

11. A coin is tossed three times. Calculate the probability of obtaining
    a) exactly one head,   b) at least one head.

12. Student A is not very bright. The probability that he will fail any course is 0.1. He has registered for five courses this semester. What is the probability that he will pass all of them?

13. An unbalanced coin has $P(\text{heads}) = 0.6$ and $P(\text{tails}) = 0.4$. When five of these coins are tossed, what is the probability of obtaining
    a) exactly two heads?   b) at least two heads.?

132   RANDOM VARIABLES

14. The characteristic function of the random variable X is given by

$$\phi_X(j\nu) = \frac{a}{a + j\nu}.$$

What is its mean?

15. Show that if a density function is symmetric about zero, the corresponding characteristic function is real and symmetric.

16. The random variable X has the characteristic function

$$\phi_X(\nu) = K \frac{\sin \nu}{\nu}.$$

a) Evaluate the constant K.
b) Calculate the mean of X.
c) Find the pdf $f_X(X)$.
d) Given that $Y = X + a$, where a is a constant, find $\phi_Y(\nu)$.

17. Suppose that the volume on your radio is turned up so that the audio amplifier is delivering 0.5 W to a one-ohm speaker.

a) Plot roughly to scale the pdf of the speaker voltage at an arbitrary time t_1.
b) Find the probability that the speaker voltage will exceed 0.2 V at time t_1.

18. Computer assignment. Write a FORTRAN program to compute the $(n + 1)$ binomial probabilities

$$P\{X = k\} = \binom{n}{k} p^k q^{(n-k)}$$

for any n and p. Thus the input data consists of two numbers, n and p, and the output is the $(n + 1)$ probabilities $P\{X = k\}$ for $k = 0, 1, \ldots, n$.

19. Computer assignment. The subroutine library for most computers contains a random number generator that produces numbers uniformly distributed $(0, 1)$.

a) Find an empirical cdf for 200 numbers $\{x_i\}$ obtained from this uniform random number generator.
b) Find an empirical cdf for 200 numbers $\{y_i\}$ obtained as follows:

$$y_1 = \sum_{i=1}^{12} x_i - 6.0, \qquad y_2 = \sum_{i=13}^{24} x_i - 6.0, \qquad \ldots,$$

where each x_i is obtained from the uniform number generator.

FURTHER READING

1. P. L. MEYER, *Introductory Probability and Statistical Applications*, Addison-Wesley, Reading, Mass., 1965.
2. A. PAPOULIS, *Probability, Random Variables, and Stochastic Processes*, McGraw-Hill, New York, 1965.
3. J. M. WOZENCRAFT and I. M. JACOBS, *Principles of Communication Engineering*, Wiley, New York, 1965.

These three texts discuss the material in this chapter with varying degrees of thoroughness. Meyer does not discuss characteristic functions because his text is written for students with little background in the use of complex functions, but Meyer and Papoulis both stress the concept of function in connection with random variables. Both 2 and 3 provide a thorough discussion of characteristic functions.

7
RANDOM VARIABLES
(*Continued*)

7.1 FUNCTIONS OF A RANDOM VARIABLE

We introduce the concept of a composite function in the following way. Suppose we are given the function f with domain D_f and range R_f. We say that f is a mapping from D_f to R_f. Suppose we are also given the function g with domain $D_g = R_f$ (the range of f) and range R_g. Figure 7.1 illustrates this concept. Then we can define a new function $y = gf = g(f)$, which is a mapping from D_f to R_g. That is, D_f is the domain of y and R_g is the range of y.

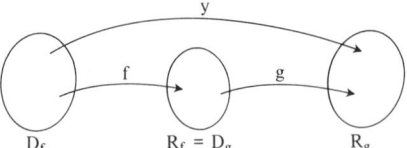

Fig. 7.1

Notes. a) In the above discussion we restricted the range of the first function, R_f, to the domain of the second, D_g. More generally, it is necessary that $R_f \subset D_g$. Thus R_f can be equal to D_g or R_f can be a proper subset of (included in but not equal to) D_g.

b) Note that f was the first function and g the second, yet in the notation gf they are turned around. It will help to keep the order of the functions straight to read gf from right to left: First apply f, then g.

For example, let $f(x) = x^2$ and $g(x) = 2x + 1$, where D_f and D_g are the set of real numbers. Then the composite function $y = g(f(x))$ is

$$y(x) = g(x^2) = 2x^2 + 1.$$

Alternatively, the composite function $v = f(g(x))$ is

$$v(x) = f(2x + 1) = (2x + 1)^2 = 4x^2 + 4x + 1.$$

Note that $gf \neq fg$.

7.2 DETERMINATION OF THE cdf AND pdf OF $Y = g(X)$

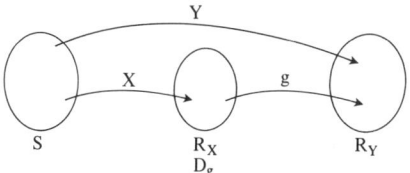

Fig. 7.2

Remember that a random variable is a function. The domain of the random variable is the sample space S and the range is some set of numbers. Label the random variable X and the range R_X. (In this text we are considering only real-valued random variables, so R_X is a set of real numbers.) According to our above discussion, any function g with domain $D_g \supset R_X$ can be a function of the random variable X, $g(X)$. The domain of the composite function $y = g(X)$ is the sample space S, and therefore can be correctly labeled a random variable if the conditions of Definition 1 in Chapter 6 are satisfied. For this reason we write $Y = g(X)$, using an upper-case Y in place of the lower-case y. Figure 7.2 illustrates this concept.

7.2 DETERMINATION OF THE cdf AND pdf OF $Y = g(X)$

Consider the following problem: Suppose we know the cdf (and hence the pdf) of the random variable X, and suppose there is a relationship of the form $Y = g(X)$. Find the cdf (and pdf) of the random variable Y.

The reader can appreciate that the solution to this problem has wide application in systems theory. The application is direct for systems without memory, for in this case the input x and the output y are related by a formula of the form $y = g(x)$. If the input is a random quantity with known cdf, then the cdf for the output can be computed. The problem is much more complicated for systems with memory, but the present discussion will form a first step toward the solution of this problem.

This problem is related to the concept of composite function as follows: Let A be a subset of the sample space, $A \subset S$. Then A is an event. Suppose that the experiment is performed and that the outcome ζ is an element of A, $\zeta \in A$. The point ζ is mapped into the number $X(\zeta)$ by the random variable X, and if there is a relation of the form $Y = g(X)$, then the number $X(\zeta)$ is mapped into $g(X(\zeta))$. (See Fig. 7.3.)

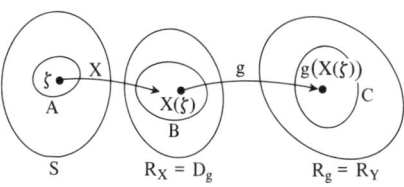

Fig. 7.3

Each element ζ of A is mapped into some element $X(\zeta)$ of B by the random variable X, which is in turn mapped into some element $g(X(\zeta))$ of C by the function g. With this in mind, let us define the important concept of equivalent events.

Definition 1. *Equivalent events.* Let A be a subset of S, and let B be a subset of R_X. Then A is equivalent to B if

$$A = \{\zeta \in S : X(\zeta) \in B\}. \tag{7.1}$$

In words, A is the set of elements ζ in S such that $X(\zeta)$ is an element of B.

Notes. a) The sets B and C in Fig. 7.3 are equivalent if

$$B = \{b \in R_X : g(b) \in C\}.$$

b) The sets A, B, and C occur together. That is, the occurrence of an element in one of the three sets implies the occurrence of an element in each of the other two sets.

c) This concept of equivalent events should not be confused with the more common (in mathematics) concept of equivalent sets as defined in Definition 6 of Chapter 5.

Two sets are equivalent if their elements can be put into one-to-one correspondence. That is, the correspondence between the two sets is arbitrary, but it must be one-to-one.

In contrast, in our definition of equivalent events the correspondence between two sets (say R_X and R_g) is specified by the function (g in this case).

Since we will have no use for the concept of equivalent sets in the following discussion, when we speak (rather loosely) of equivalent events and/or equivalent sets, we will always mean the concept given in Definition 1.

d) Recall that an event is a set. That is, an event is a subset of the sample space S. What we have done here is to find the corresponding set in the range R_X. We can now think of the corresponding set B (equivalent to A) as an event, and of the range R_X as a sample space. We will find this convenient later, for R_X is a set of numbers and is inherently easier to work with than the sample space S.

Important Property

Equivalent events are important because their probabilities are equal. Let A and B be equivalent events. Then $P(A) = P(B)$.

Notes. a) This property is important because it is the basis of the procedure for finding the cdf of $Y = g(X)$ given below.

b) It is important to keep in mind that A and B are in different sets. In Definition 1, A is a subset of the sample space S and B is a subset of the range R_X.

7.2 DETERMINATION OF THE cdf AND pdf OF $Y = g(X)$

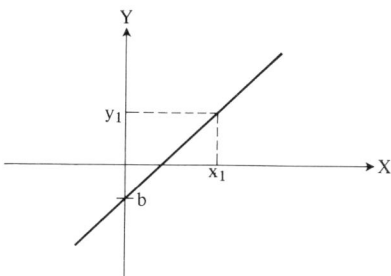

Fig. 7.4

Finding the cdf of $Y = g(X)$

We may now easily (logically) deduce the cdf of $Y = g(X)$ if we know the cdf of X. For a particular value of $y = y_1$ the function F_Y defines an event in R_Y, namely $\{Y \leq y_1\}$. There is an equivalent event in R_X, call it B. Therefore

$$F_Y(y_1) = P\{Y \leq y_1\} = P\{X \in B\}, \qquad (7.2)$$

and we can express the probability $P\{X \in B\}$ in terms of F_X.

The process of finding $P\{X \in B\}$ in terms of F_X deserves some discussion. Then we will illustrate this process with several examples (Section 7.3).

To begin with a simple situation, suppose g is a straight-line relationship of the form $ax + b$. That is,

$$Y = aX + b. \qquad (7.3)$$

There are two cases to consider (because they lead to different results): $a > 0$ and $a < 0$.

First consider the case $a > 0$ (Fig. 7.4). The events $\{Y \leq y_1\}$ and $\{X \leq x_1\}$ are equivalent, and therefore

$$F_Y(y_1) = P\{Y \leq y_1\} = P\{X \leq x_1\} = F_X(x_1).$$

But $x_1 = (y_1 - b)/a$, so

$$F_Y(y_1) = F_X\left(\frac{y_1 - b}{a}\right).$$

Therefore, for any y the relationship between F_Y and F_X is

$$F_Y(y) = F_X\left(\frac{y - b}{a}\right). \qquad (7.4)$$

As an example, suppose X is uniform $(3, 5)$ and Y is related to X by $Y = 2X + 4$. Then

$$F_X(x) = \begin{cases} 0, & x < 3, \\ \tfrac{1}{2}(x - 3), & 3 < x < 5, \\ 1, & x > 5, \end{cases}$$

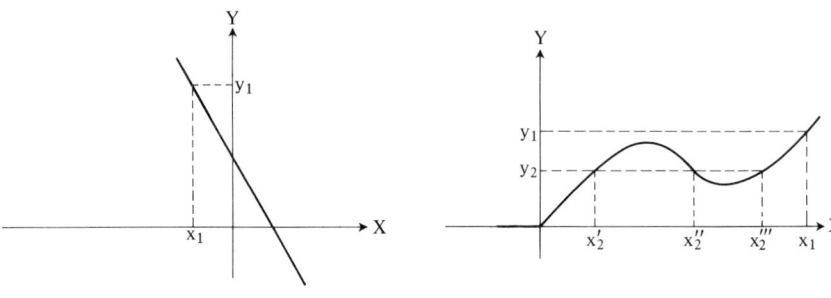

Fig. 7.5 Fig. 7.6

so, from Eq. (7.4), F_Y is given by

$$F_Y(y) = \begin{cases} 0, & y < 10, \\ \frac{1}{4}(y - 10), & 10 < y < 14, \\ 1, & y > 14. \end{cases}$$

Next consider the case $a < 0$ (Fig. 7.5). Now the events $\{Y \leq y_1\}$ and $\{X \geq x_1\}$ are equivalent. Therefore, $F_Y(y_1) = P\{Y \leq y_1\} = P\{X \geq x_1\} = 1 - F_X(x_1)$. Again $x_1 = (y_1 - b)/a$, so

$$F_X(y_1) = 1 - F_X\left(\frac{y_1 - b}{a}\right),$$

and for any y the relationship between F_Y and F_X is

$$F_Y(y) = 1 - F_X\left(\frac{y - b}{a}\right). \tag{7.5}$$

As an example, again suppose X is uniform $(3, 5)$ and Y is related to X by $Y = -2X + 4$. By Eq. (7.5), F_Y is given by

$$F_Y(y) = \begin{cases} 0, & y < -6, \\ \frac{1}{4}(y + 6), & -6 < y < -2, \\ 1, & y > -2. \end{cases}$$

Now consider a more general (and more complicated) case. Instead of the straight-line relationship (Eq. 7.3), suppose $y = g(x)$ is the curve shown in Fig. 7.6.

For $y = y_1$ the sets $\{Y \leq y_1\}$ and $\{X \leq x_1\}$ are equivalent. Therefore

$$F_Y(y_1) = P\{Y \leq y_1\} = P\{X \leq x_1\} = F_X(x_1). \tag{7.6}$$

For $y = y_2$ there are three solutions to $y = g(x)$, namely, x_2', x_2'', and x_2'''. From the graph we see that the equivalent events are

$$\{Y \leq y_2\} \sim \{x_2'' < X \leq x_2'''\} \cup \{X \leq x_2'\}. \tag{7.7}$$

7.3 FINDING THE DISTRIBUTION OF $Y = g(X)$

(The symbol $\sim$ denotes equivalence.) Hence,

$$F_Y(y_2) = P\{Y \leq y_2\} = P\{x_2'' < X \leq x_2'''\} + P\{X \leq x_2'\}$$
$$= F_X(x_2''') - F_X(x_2'') + F_X(x_2'). \tag{7.8}$$

Finally, for $y < 0$ (say $y = y_3$) there is no solution to the equation $y_3 = g(x)$. This means that there is no event in R_X equivalent to $\{Y \leq y_3\}$ in R_Y. Therefore

$$F_Y(y_3) = P\{Y \leq y_3\} = 0. \tag{7.9}$$

Notes. a) The fact that equivalent events have equal probabilities is the basis for the above procedure.

b) This procedure is quite general and also straightforward. Given the cdf of the random variable Y we can find the pdf. Now there are procedures for finding the pdf f_Y from the pdf f_X directly, and in the examples that follow we will indicate just what this procedure is. Generally speaking, however, we can avoid much unnecessary complication by working with the cdf, as above.

7.3 EXAMPLES OF FINDING THE DISTRIBUTION OF $Y = g(X)$

1. Square-law device. Suppose that a system has the input-output characteristic $y = x^2$ shown in Fig. 7.7(b) and that the input $x(t)$ is the periodic sawtooth wave shown in Fig. 7.7(a). If the voltage $x(t)$ is sampled at some arbitrary time t, then we can define a random variable X whose range is the possible values of $x(t)$, that is, whose range is the set of numbers $[0, a]$. The random variable X is uniformly distributed $(0, a)$. Find F_Y and f_Y.

Let $y = y_1$, where y_1 is any number greater than zero, as shown in Fig. 7.7(b). Then the equivalent events are

$$\{Y \leq y_1\} \sim \{x_1' \leq X \leq x_1''\}, \tag{7.10}$$

where x_1' and x_1'' are solutions to the equation $y = x^2$. That is,

$$x_1' = -\sqrt{y_1} \quad \text{and} \quad x_1'' = \sqrt{y_1}. \tag{7.11}$$

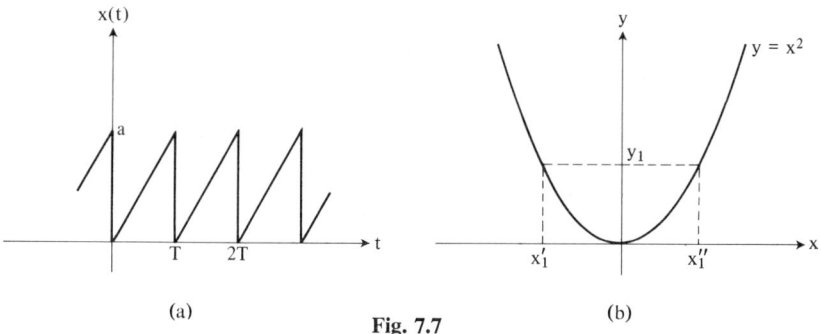

Fig. 7.7

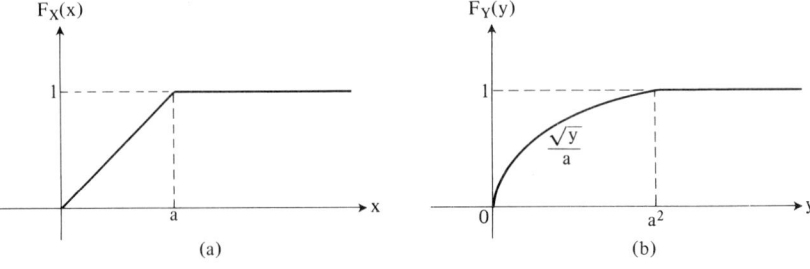

Fig. 7.8

Then
$$F_Y(y_1) = P\{Y \le y_1\} = P\{-\sqrt{y_1} < X < \sqrt{y_1}\}$$
$$= F_X(\sqrt{y_1}) - F_X(-\sqrt{y_1}), \quad y_1 > 0. \quad (7.12)$$

The cdf of X is shown in Fig. 7.8(a) and is given by
$$F_X(x) = \begin{cases} 0, & x < 0, \\ \dfrac{1}{a}x, & 0 < x < a, \\ 1, & x > a. \end{cases} \quad (7.13)$$

Substituting this into Eq. (7.12), we get the cdf of Y:
$$F_Y(y) = \begin{cases} 0, & y < 0, \\ \dfrac{1}{a}\sqrt{y}, & 0 < \sqrt{y} < a \text{ or } 0 < y < a^2, \\ 1, & y > a^2. \end{cases} \quad (7.14)$$

We can find the corresponding pdf by differentiating Eq. (7.14) with respect to y. Thus we get
$$f_Y(y) = \begin{cases} 0, & y < 0, \\ \dfrac{1}{2a\sqrt{y}}, & 0 < y < a^2, \\ 0, & y > a^2. \end{cases} \quad (7.15)$$

Note. We could have found the pdf f_Y directly from the pdf f_X without using the cdf's $F_X(x)$ and $F_Y(y)$. The procedure is as follows: Let $y = g(x)$ be the relationship between y and x. Then
$$f_Y(y) = \frac{f_X(x_1)}{|g'(x_1)|} + \frac{f_X(x_2)}{|g'(x_2)|} + \cdots + \frac{f_X(x_n)}{|g'(x_n)|}, \quad (7.16)$$

7.3 FINDING THE DISTRIBUTION OF $Y = g(X)$

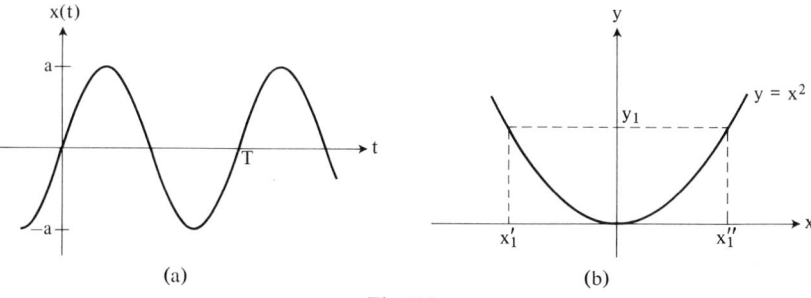

Fig. 7.9

where $x_1, x_2, \ldots, x_n$ are all real roots of $y = g(x)$ and

$$g'(x) = \frac{dg(x)}{dx}. \tag{7.17}$$

Obviously this procedure is applicable only when there are a finite number of roots of $y = g(x)$. For a thorough discussion of this procedure, see Chapter 5 of reference 3.

2. Now suppose the sinusoidal waveform of Fig. 7.9(a) is supplied to the square-law device. Again we define the random variable X so that the range of X is all possible values of $x(t)$. The distribution of the random variable X is that of Example 6 of Section 6.3. We will now derive this distribution.

Sampling $x(t)$ at some specific but arbitrary time t is analogous to assuming that $x(t)$ is given by

$$x(t) = a \sin(\omega t + \Theta), \tag{7.18}$$

where Θ is a random variable uniformly distributed in the interval $(0, 2\pi)$ and t is a fixed parameter. The relationship between the random variables X and Θ is therefore given by

$$X = a \sin \Theta, \quad 0 < \Theta < 2\pi. \tag{7.19}$$

The cdf of Θ is shown in Fig. 7.10(a) and the relationship between X and Θ is shown in Fig. 7.10(b). We now wish to find F_X in terms of F_Θ.

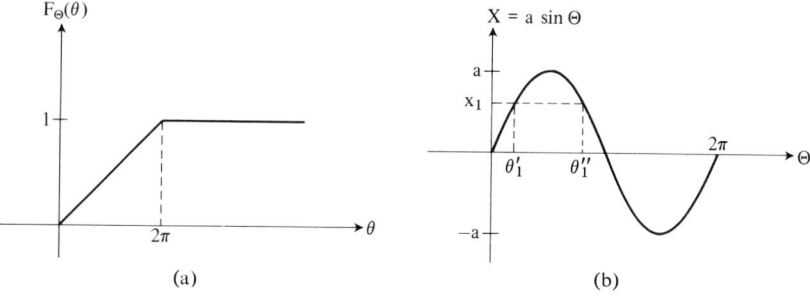

Fig. 7.10

For $x > a$ the event $\{X \leq x\}$ is certain. Therefore

$$F_X(x) = 1, \quad x > a. \tag{7.20}$$

For $x = x_1$, where $0 < x_1 < a$ (see Fig. 7.10b), there are two solutions to the equation $x_1 = a \sin \theta$, namely, θ_1' and θ_1'', as shown in Fig. 7.10(b). The equivalent events are

$$\{X \leq x_1\} \sim \{\Theta \leq \theta_1'\} \cup \{\theta_1'' \leq \Theta\} \tag{7.21}$$

and

$$F_X(x_1) = P\{X \leq x_1\} = P\{\Theta \leq \theta_1'\} + P\{\theta_1'' \leq \Theta\}$$

$$= F_\Theta(\theta_1') + 1 - F_\Theta(\theta_1''). \tag{7.22}$$

The two solutions θ_1' and θ_1'' are given by

$$\theta_1' = \sin^{-1}\frac{x_1}{a} \quad \text{and} \quad \theta_1'' = \pi - \sin^{-1}\frac{x_1}{a}. \tag{7.23}$$

Substituting these values into Eq. (7.22) (where F_Θ is as given in Fig. 7.10) yields

$$F_X(x_1) = \frac{1}{2} + \frac{1}{\pi}\sin^{-1}\frac{x_1}{a}, \quad 0 < x_1 < a. \tag{7.24}$$

Now suppose $x = x_2$, where $-a < x_2 < 0$. The equivalent events are

$$\{X \leq x_2\} \sim \{\theta_2' \leq \Theta \leq \theta_2''\}. \tag{7.25}$$

By a procedure identical to that given above we arrive at the conclusion that $F_X(x_2)$ is again given by Eq. (7.24) for $-a < x_2 < 0$.

Finally, for $x < -a$, $F_X(x) = 0$. Figure 7.11 illustrates the cdf and pdf of X, where the pdf is found by differentiating the cdf with respect to x.

If you've followed the discussion this far, the solution to our original problem should be trivial—or at least easy, compared to the preceding problem.

For $y = y_1$, where $y_1 > 0$ (see Fig. 7.9), the equivalent events in R_Y and R_X are

$$\{Y \leq y_1\} \sim \{x_1' \leq X \leq x_1''\}. \tag{7.26}$$

Therefore

$$F_Y(y_1) = P\{Y \leq y_1\} = P\{x_1' \leq X \leq x_1''\}$$

$$= F_X(x_1'') - F_X(x_1') \tag{7.27}$$

$$= \frac{2}{\pi}\sin^{-1}\frac{\sqrt{y_1}}{a}, \quad 0 < y_1 < a^2. \tag{7.28}$$

3. Limiter. Suppose that the relationship between y and x is that shown in Fig. 7.12, rather than the square-law relationship of the previous two examples. The input to the limiter is a sine wave of amplitude a. Find the cdf and pdf of the output $y(t)$.

7.3 FINDING THE DISTRIBUTION OF $Y = g(X)$

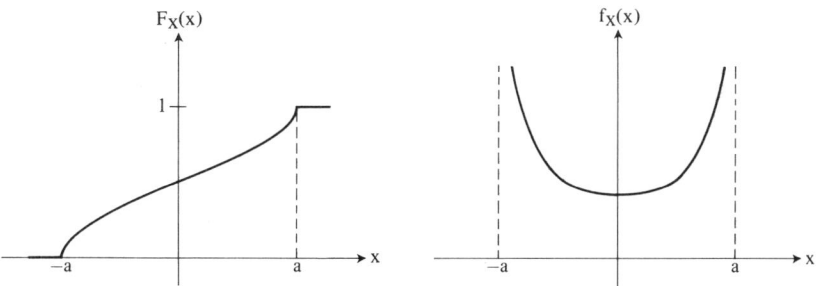

Fig. 7.11

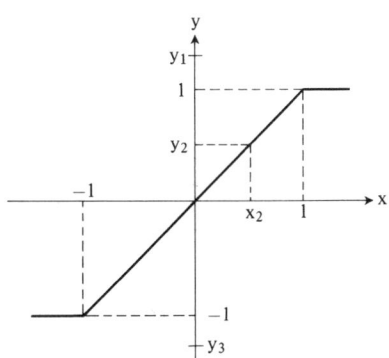

Fig. 7.12

First, for $a < 1$ the limiter has no effect on the input $x(t)$ and the cdf and pdf are as shown in Fig. 7.11. This is the trivial case, and is of no interest.

Suppose $a > 1$. For $y = y_1$, where $y_1 > 1$, the event $\{Y \leq y_1\}$ is certain. Hence

$$F_Y(y_1) = P\{Y \leq y_1\} = 1, \quad y_1 > 1. \quad (7.29)$$

For $y = y_2$, where $-1 < y_2 < 1$ (see Fig. 7.12), the equivalent events in R_Y and R_X are given by

$$\{Y \leq y_2\} \sim \{X \leq x_2\}. \quad (7.30)$$

Therefore

$$F_Y(y_2) = P\{Y \leq y_2\} = P\{X \leq x_2\} = F_X(x_2). \quad (7.31)$$

Finally, for $y = y_3$, where $y_3 < -1$, the event $\{Y \leq y_3\}$ can never occur. The equivalent event in R_X is the empty set. The cdf of Y is

$$F_Y(y) = \begin{cases} 1, & y > 1, \\ \dfrac{1}{2} + \dfrac{1}{\pi} \sin^{-1} \dfrac{y}{a}, & -1 < y < 1, \\ 0, & y < -1. \end{cases} \quad (7.32)$$

This cdf and the pdf are shown in Fig. 7.13.

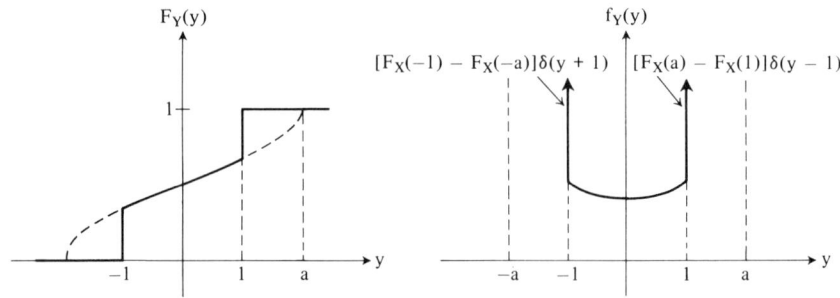

Fig. 7.13

Notes. a) This distribution is of mixed type (both continuous and discrete). There are jumps in the cdf and impulses in the pdf.

b) The procedure for solving these problems is always the same: Find an event in R_X that is equivalent to the event $\{Y \leq y\}$ in R_Y. Once this equivalent event has been determined, the problem is straightforward.

7.4 JOINTLY DISTRIBUTED RANDOM VARIABLES

Consider performing an experiment where events are defined on a sample space and probabilities are assigned to these events. Suppose there are two random variables X and Y defined on the sample space. (See Fig. 7.14.) If the experimental outcome is ζ, then two numbers are determined: $X(\zeta)$ and $Y(\zeta)$. The ordered pair (X, Y) is called a *two-dimensional random variable* or a *random vector*.

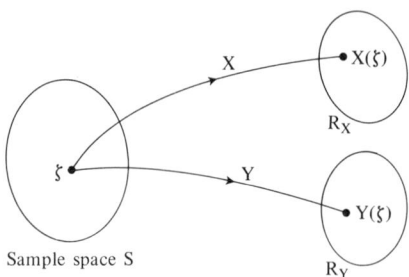

Fig. 7.14 Graphic illustration of a two-dimensional random variable.

Notes. a) There is nothing random about a two-dimensional random variable, just as there is nothing random about a one-dimensional random variable. A two-dimensional random variable is a function. The randomness occurs in the experiment.

b) The experiment is performed *once* and an ordered pair of numbers is generated by (X, Y). This is quite different from performing the experiment twice and using a single random variable X to generate the ordered pair $(X(\zeta_1), X(\zeta_2))$.

The term "joint random variable" or "jointly distributed random variable" is applied to a two-dimensional, three-dimensional, or higher-dimensional random variable. A three-dimensional random variable consists of the 3-tuple generated by one experimental outcome, say $(X(\zeta), Y(\zeta), Z(\zeta))$. Thus an illustration similar to Fig. 7.14 for the three-dimensional random variable would consist of three range spaces, R_X, R_Y, and R_Z, and the one sample space S.

Example 4. Let us apply this concept to sample values taken at times t_1, $t_2, \ldots, t_k$. Consider any waveform. The one given in Fig. 6.10 will do. Suppose we take k samples spaced one second apart. Thus we have k values $v(t_1), v(t_2), \ldots, v(t_k)$, and if the time of the first sample t_1 is arbitrary, we have a k-dimensional random variable.

Notes. a) In this example the experimental outcome ζ determines the sampling time t_1 (and therefore all other sampling times). The domain or sample space for our random variable is a set of experimental outcomes (about which we have said nothing), and one element of the range is the k sample values $v(t_1), v(t_2), \ldots, v(t_k)$.

b) In this example we have chosen to let the experimental outcome determine the sampling times. An alternative procedure for choosing a random vector is to fix the sampling times and to let the experimental outcome determine different voltage waveforms. We will discuss this procedure in the chapter on stochastic processes, Chapter 9.

c) An alternative name for an n-dimensional random variable is a random vector.

Definition 2. *Joint cdf.* The joint cumulative distribution function F_{XY} is the probability

$$F_{XY}(x, y) = P\{X \leq x, Y \leq y\}. \tag{7.33}$$

Notes. a) This is the joint probability. A more accurate way to write it is $P(\{X \leq x\} \cap \{Y \leq y\})$. A comma is commonly used throughout the literature in place of the intersection symbol.

b) The subscripts on F_{XY} distinguish this particular function from other (different) functions that might be labeled F.

c) The domain of F_{XY} is the set of all ordered pairs (x, y), where $x \in R_X$ and $y \in R_Y$ (see Fig. 7.14). The range of F_{XY} is the set of numbers $[0, 1]$.

d) $F_{XY}(x, y)$ is the probability of the event $\{X \leq x, Y \leq y\}$. There must be an equivalent event in S with probability equal to that of $F_{XY}(x, y)$. We will make use of this equivalent event later.

e) This concept is readily extended to higher dimensions (Eq. 7.33 defines the two-dimensional cdf). For a three-dimensional random variable the joint cdf is

$$F_{XYZ}(x, y, z) = P\{X \leq x, Y \leq y, Z \leq z\}.$$

Definition 3. *Joint pdf.* The joint probability density function f_{XY} is

$$f_{XY}(x, y) = \frac{\partial^2}{\partial x \, \partial y} F_{XY}(x, y). \tag{7.34}$$

Notes. a) For discrete random variables the joint pdf will have impulses in the two-dimensional plane. For continuous random variables the joint pdf will be a smooth surface.

b) As before, the pdf is a *density* function and hence we must integrate it to find probability. For the two-dimensional random variable we must integrate the pdf twice to obtain probability. For example, the probability that (X, Y) is in the rectangle $(x_1 \leq X \leq x_2)$ and $(y_1 \leq Y \leq y_2)$ is given by

$$P\{x_1 \leq X \leq x_2, y_1 \leq Y \leq y_2\} = \int_{x_1}^{x_2} \int_{y_1}^{y_2} f_{XY}(x, y) \, dy \, dx. \tag{7.35}$$

c) Again this concept applies to higher dimensions.

We now provide an example of a two-dimensional random variable and calculate the cdf and pdf.

Example 5. Suppose both X and Y are discrete, and suppose our experiment consists of tossing two coins. The experimental outcome is the number of heads that turn up when the two coins are tossed. The three possible outcomes are

ζ_1 = no heads turn up,
ζ_2 = one head turns up,
ζ_3 = two heads turn up.

The joint random variable (X, Y) is defined in Fig. 7.15(a). Figure 7.15(b) and (c) illustrate the pdf and cdf, respectively.

Note. In this example both X and Y are discrete, so the pdf consists of impulses in the plane and the cdf consists of steps or jumps in both the x- and y-directions. In the next chapter, Fig. 8.3 illustrates some two-dimensional continuous pdf's. The associated cdf's are not shown, but they are smooth surfaces that increase in height in both the positive x- and y-directions (to a maximum height of one). It is also possible to have one discrete random vari-

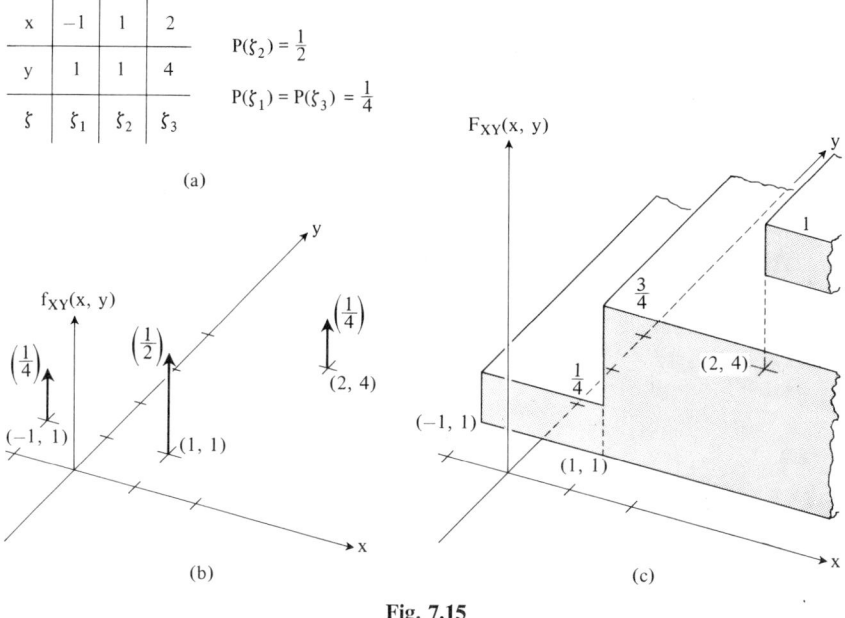

Fig. 7.15

able and one continuous random variable. In this case the two-dimensional pdf would consist of "fences" in the plane, and the cdf would have steps in one direction but not in the other.

Definition 4. *Independent random variables.* Two random variables X and Y are independent if the events $\{X < x\}$ and $\{Y < y\}$ are statistically independent for all values of x and y.

The independence of X and Y implies the following:

1. $F_{XY}(x, y) = F_X(x)F_Y(y)$.
2. $f_{XY}(x, y) = f_X(x)f_Y(y)$.
3. $E(XY) = E(X)E(Y)$.

The converse of statement 3 is not true. If $E(XY) = E(X)E(Y)$, the random variables are said to be uncorrelated. Random variables can be uncorrelated but not independent. The converses of 1 and 2 are true. That is, if $f_X(x)f_Y(y) = f_{XY}(x, y)$, then the random variables X and Y are independent. Also, if $F_{XY}(x, y) = F_X(x)F_Y(y)$, then X and Y are independent.

In dealing with two or more jointly distributed random variables, we call the distribution of each random variable the *marginal distribution*. We define this marginal distribution in terms of the joint distribution and then show that this does indeed result in the distribution of each random variable considered by itself.

Definition 5. *Marginal distribution.* $F_X(x)$, in terms of $F_{XY}(x, y)$, is

$$F_X(x) = F_{XY}(x, \infty). \tag{7.36}$$

Similarly,

$$F_Y(y) = F_{XY}(\infty, y). \tag{7.37}$$

To see that Eq. (7.36) defines the distribution of X, note that the events $\{X \leq x, Y \leq \infty\}$ and $\{X \leq x\}$ are equivalent; that is,

$$\{X \leq x, Y \leq \infty\} \sim \{X \leq x\}. \tag{7.38}$$

Hence their probabilities are equal.

From this we can show that the marginal density functions $f_X(x)$ and $f_Y(y)$ are related to the joint density function $f_{XY}(x, y)$. From Eq. (7.36) we have

$$F_X(x) = F_{XY}(x, \infty) = \int_{-\infty}^{+\infty} \int_{-\infty}^{x} f_{XY}(\lambda, y) \, d\lambda \, dy. \tag{7.39}$$

Differentiating both sides with respect to x, we get

$$f_X(x) = \int_{-\infty}^{+\infty} f_{XY}(x, y) \, dy. \tag{7.40}$$

Similarly,

$$f_Y(y) = \int_{-\infty}^{+\infty} f_{XY}(x, y) \, dx. \tag{7.41}$$

Note. Given the joint distribution (density) function, we can find the marginal distribution (density) functions. We cannot, in general, find the joint function if the marginal functions are known. An exception occurs when X and Y are independent. Then the joint function is the product of the marginal functions.

Joint Moments

Definition 6. *Joint moments.* The joint moment m_{ij} is the expected value of the product $X^i Y^j$. It is given by

$$m_{ij} = E(X^i Y^j) = \int_{-\infty}^{+\infty} \int_{-\infty}^{+\infty} x^i y^j f_{XY}(x, y) \, dx \, dy. \tag{7.42}$$

Note. The joint moment is the average value of the product of the values of the random variables X^i and Y^j. This is not to be confused with composite functions.

The order of the moment is $i + j$. There are three second-order moments, m_{20}, m_{11}, m_{02}, and these all have special significance. Obviously $m_{20} = E(X^2)$ and $m_{02} = E(Y^2)$, the mean-square values of X and Y, respectively. The moment m_{11} is of such significance that some special terminology has been developed for it. The special significance of m_{11} will be demonstrated in our discussion of stochastic processes. For the time being, we will simply introduce the special terminology.

7.5 CONDITIONAL PROBABILITY DISTRIBUTION

The second central moment μ_{11} is called the *covariance* of X and Y. It is given by

$$\mu_{11} = E[(X - m_X)(Y - m_Y)]$$
$$= E(XY) - E(x)E(y) = m_{11} - m_X m_Y. \quad (7.43)$$

The *correlation coefficient* ρ_{XY} is

$$\rho_{XY} = \frac{\mu_{11}}{\sigma_X \sigma_Y}, \quad (7.44)$$

with the important property that

$$-1 \leq \rho_{XY} \leq 1. \quad (7.45)$$

Note. If $\rho_{XY} = 0$, the random variables X and Y are uncorrelated. Refer to property 3 on page 147 and Eq. (7.43).

Example 6. Refer to Example 5, with the joint random variables X and Y as defined in Fig. 7.15(a). We now calculate the second moment m_{11}, the second central moment μ_{11}, and the correlation coefficient ρ_{XY}:

$$m_{11} = E(XY) = \sum_{x,y} xy f_{XY}(x, y) = (-1)(1)\tfrac{1}{4} + (1)(1)\tfrac{1}{2} + (2)(4)\tfrac{1}{4} = 2\tfrac{1}{4},$$

$$\mu_{11} = E[(X - m_X)(Y - m_Y)] = \sum_{x,y}(x - m_X)(y - m_Y) f_{XY}(x, y)$$

$$= (-1 - \tfrac{3}{4})(1 - \tfrac{7}{4})\tfrac{1}{4} + (1 - \tfrac{3}{4})(1 - \tfrac{7}{4})\tfrac{1}{2} + (2 - \tfrac{3}{4})(4 - \tfrac{7}{4})\tfrac{1}{4} = \tfrac{15}{16},$$

$$\rho_{XY} = \frac{\mu_{11}}{\sigma_X \sigma_Y} = \frac{\tfrac{15}{16}}{\sqrt{\tfrac{19}{16}\tfrac{27}{16}}} \approx 0.663.$$

Note that according to our discussing in Section 2.2, random variables can be thought of as vectors. Let us define the inner product as $E(XY)$. Then $E(XY) = 0$ implies that X and Y are orthogonal. This concept of orthogonality is quite useful in mean-square estimation, and we will find uses for it in Chapter 11.

7.5 CONDITIONAL PROBABILITY DISTRIBUTION

In Section 5.5 we introduced the concept of conditional probability and we stressed the point that a reduction in the sample space results from the conditioning. We now extend this concept to random variables and consider the meaning of expressions such as $f_X(x|y)$ and $F_X(x|y)$. Here we wish to draw an analogy between conditional pdf (cdf) and conditional probability.

First consider expressions such as $F_X(x|A)$. This is the cdf for the random value X, given that the event A has occurred. Since A is a subset of the sample space (by definition), the possible values of X are restricted to those related to experimental outcomes $\zeta \in A$. (See Fig. 7.16.) An example should clarify these concepts.

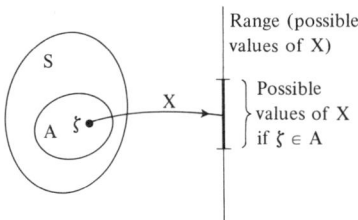

Fig. 7.16

Examples

7. A single die is thrown and X is the usual random variable for this experiment. That is, if a 1 is rolled then $X = 1$; if a 2 is rolled then $X = 2$, etc. The cdf $F_X(x)$ is plotted in Fig. 7.17(a). (This is the unconditional cdf.)

Let A be the event "X is greater than 3." The conditional cdf $F_X(x|A)$ is plotted in Fig. 7.17(b). In the unconditional case X can take on all possible values $\{1, 2, 3, 4, 5, 6\}$. In the conditional case the sample space is restricted to 4, 5, and 6, so X can take on only the values $\{4, 5, 6\}$.

Another viewpoint is possible. Recall that conditional probability is related to joint and unconditional probability by

$$P(B|A) = \frac{P(B \cap A)}{P(A)}.$$

Since F_X is a probability, we can write

$$F_X(x|A) = P\{X \leq x|A\} = \frac{P\{X \leq x \cap A\}}{P(A)}.$$

8. Consider again the situation in Example 7. The event A has probability $\frac{1}{2}$ and

$$P\{X \leq 4 \cap A\} = P\{X = 4\} = \tfrac{1}{6},$$
$$P\{X \leq 5 \cap A\} = P\{X = 4 \cup X = 5\} = \tfrac{2}{6},$$
$$P\{X \leq 6 \cap A\} = P\{X = 4 \cup X = 5 \cup X = 6\} = \tfrac{3}{6}.$$

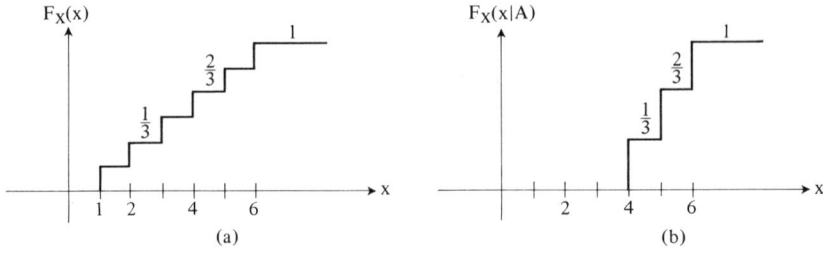

Fig. 7.17

7.5 CONDITIONAL PROBABILITY DISTRIBUTION 151

X	-2	-1	0	1	2	3
Y	4	1	0	1	4	9
ζ	⚀	⚁	⚂	⚃	⚄	⚅

Fig. 7.18

Hence
$$P\{X \leq 4|A\} = \tfrac{1}{6} \cdot 2 = \tfrac{1}{3},$$
$$P\{X \leq 5|A\} = \tfrac{2}{6} \cdot 2 = \tfrac{2}{3},$$
$$P\{X \leq 6|A\} = \tfrac{3}{6} \cdot 2 = 1,$$

in agreement with Fig. 12.17(b).

To extend these concepts slightly, suppose there are two random variables X and Y defined on the same sample space. A simple example of this situation is shown in Fig. 7.18, where the experiment is to throw a die. The experimental outcome determines a value for X and for Y. (This example will again be used in Chapter 11.)

Now consider the meaning of the expression $F_X(x|y)$. By this we mean

$$F_X(x|y) = F_X(x|Y = y)$$
$$= P\{X \leq x|Y = y\}$$
$$= \frac{P\{X \leq x \cap Y = y\}}{P\{Y = y\}}. \quad (7.46)$$

As shown in Fig. 7.19, the cdf of the random variable X, given that $Y = 4$, is given by

$$F_X(x|Y = 4) = P\{X \leq x|Y = 4\}$$
$$= \begin{cases} 0, & x < -2, \\ \tfrac{1}{2}, & -2 \leq x < 2, \\ 1, & 2 \leq x. \end{cases}$$

Notes. a) We have discussed two situations: first, $F_X(x|A)$, where A is an event, and second, $F_X(x|y)$, where y is the value of the random variable Y that

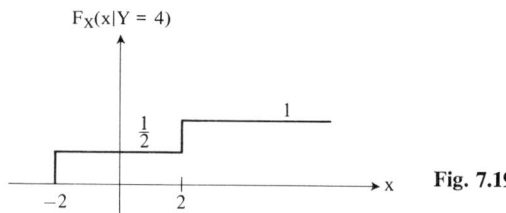

Fig. 7.19

is somehow related to X. For discrete random variables there is virtually no difference, but for continuous random variables the difference is important. See (d) below.

b) The expression $F_X(x|y)$ is often written as $F_{X|Y}(x|y)$ to differentiate this function from $F_X(x)$. However, the present notation is sufficient for our purposes.

c) The associated pdf is defined as the derivative in all cases. For instance,

$$f_X(x|y) = \frac{d}{dx} F_X(x|y).$$

d) All our discussion and examples thus far have been concerned with discrete random variables X and (especially) Y. Equation (7.46) is not valid if Y is continuous at $Y = y$ because then $P\{Y = y\} = 0$. We now discuss this situation.

Let Y be a continuous random variable at $Y = y$. Then the conditional cdf can no longer be defined by Eq. (7.46) because the denominator is zero. However, if the experiment is performed and we are told that $Y = y$, then the expressions $f_X(x|y)$ and $F_X(x|y)$ must have some meaning (and value), and the values of these functions should, in general, be different from the unconditional functions.

It is more convenient to simply define the functions $f_X(x|y)$ and $F_X(x|y)$ if Y is continuous at y, rather than relate their meaning to previous definitions as in Eq. (7.46). In doing this we wish to make our definitions consistent with all our previous definitions. For example, the total area under $f_X(x|y)$ must always be unity, just as the total area under $f_X(x)$ must be unity. Also, our definition for $F_X(x|y)$ should reduce to Eq. (7.46) if Y is discrete (discontinuous) at $Y = y$.

Definition 7. *Conditional pdf.* The conditional pdf of the random variable X, given that $Y = y$, is

$$f_X(x|y) = \frac{f_{XY}(x, y)}{f_Y(y)}, \qquad f_Y(y) > 0. \qquad (7.47)$$

Notes. a) The denominator of Eq. (7.47) is a number. For a given y, $f_Y(y)$ is just the value of f_Y evaluated at y.

b) The area under $f_X(x|y)$ is unity because

$$\int_{-\infty}^{+\infty} f_X(x|y)\, dx = \int_{-\infty}^{+\infty} \frac{f_{XY}(x, y)}{f_Y(y)}\, dx = \frac{1}{f_Y(y)} \int_{-\infty}^{+\infty} f_{XY}(x, y)\, dx = \frac{f_Y(y)}{f_Y(y)} = 1.$$

c) A geometric interpretation of the conditional pdf is given in Section 8.2. (See Fig. 8.4.)

Obviously the conditional cdf is the integral of the conditional pdf, but we will formalize this by a definition.

Definition 8. *Conditional cdf.* The conditional cdf of the random variable X is

$$F_X(x|y) = \int_{-\infty}^{x} f_X(\lambda|y) \, d\lambda. \tag{7.48}$$

Notes. a) This function satisfies the same conditions as the unconditional cdf. For example,

$$\lim_{x \to -\infty} F_X(x|y) = 0 \quad \text{and} \quad \lim_{x \to +\infty} F_X(x|y) = 1.$$

b) Suppose Y is discrete at $Y = y$. Then $f_Y(y) = P\{Y = y\}$, and Eq. (7.48) gives the same result as Eq. (7.46).

PROBLEMS

1. The periodic voltage $v(t)$ is shown in Fig. 7.20. This voltage is applied to an amplifier with gain A.

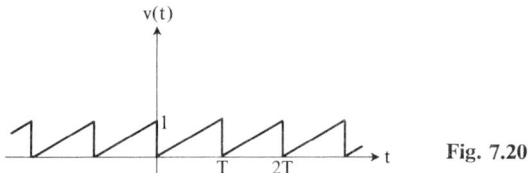

Fig. 7.20

   a) Define a random variable V whose range is the values of $v(t)$. That is, describe what is meant by the random variable V.
   b) Find the pdf and cdf of V.
   c) Find the pdf and cdf of the amplifier output.

2. If $y = ax + b$, what is the density function of y in terms of the density function of x?

3. Repeat Problem 2 for $y = ax^2 + b$.

4. A number x is randomly chosen between 0 and 1. Find $F_Y(y)$ and $f_Y(y)$ for the variables

   a) $Y = X$,   b) $Y = X^2$,   c) $Y = e^X$.

   Plot $F_Y(y)$ and $f_Y(y)$ in each case.

5. A periodic triangular waveform $v(t)$ is applied to the circuit shown in Fig. 7.21.

   a) What is the probability that a measurement of $y(t)$ is greater than 2 V?
   b) Find the mean-square value of the output $y(t)$.

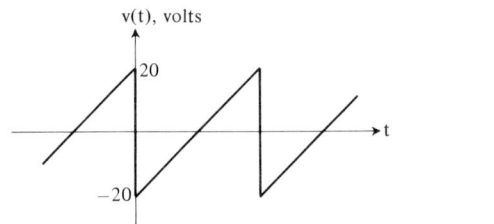

Fig. 7.21

6. Find the output cdf and pdf for each of the input-output characteristics shown in Fig. 7.22. The input X is uniformly distributed $(0, 2)$.
7. Repeat Problem 6 for X normally distributed $(0, 1)$.
8. Repeat Problem 6 for X discrete, taking on the values 1, 2, 3, 4, 5, 6 with equal probability.
9. Let $X = \tan \Theta$, where Θ is a random variable with pdf

$$f(\theta) = \begin{cases} 1/2\pi, & -\pi < \theta < \pi, \\ 0, & \text{elsewhere.} \end{cases}$$

Find the cdf and pdf of the random variable X.

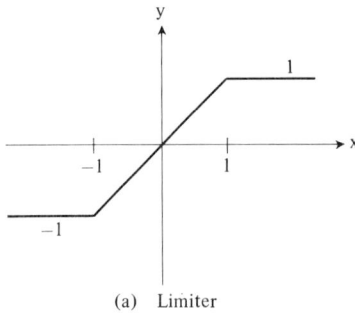

(a) Limiter

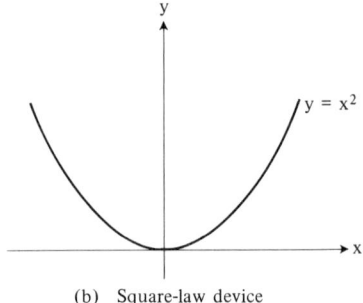

(b) Square-law device

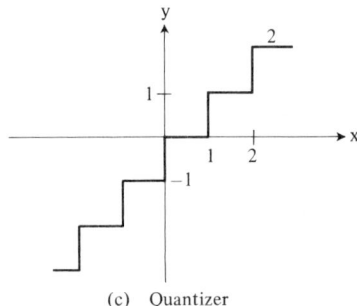

(c) Quantizer

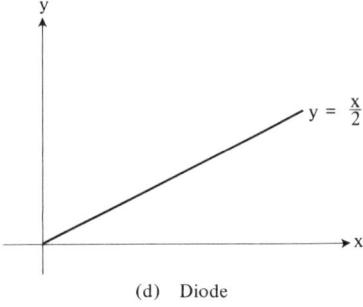

(d) Diode

Fig. 7.22

PROBLEMS 155

10. A full wave rectifier has the input-output characteristic $y = |x|$. If the input is normally distributed (0, 1), find the pdf of the output.

11. The random variables R and Θ are jointly distributed, with the pdf

$$f_{R,\Theta}(r, \theta) = \frac{r}{2\pi} e^{-r^2/2}, \quad 0 < \theta < 2\pi, \ 0 < r < \infty.$$

a) Find the probability that $\{0 < \Theta < \pi/2\}$.
b) Find the probability that $\{0 < R < \frac{1}{2}, 0 < \Theta < \pi/2\}$.

12. Find the marginal pdf's $f_R(r)$ and $f_\Theta(\theta)$ for the random variables R and Θ in Problem 11.

13. Two statistically independent random variables X_1 and X_2 have identical distribution:

$$f_X(x) = \begin{cases} \frac{1}{3}, & -1 < x < 2, \\ 0, & \text{elsewhere.} \end{cases}$$

a) Find the probability that $\{X_1 > X_2\}$.
b) Find the probability that $\{X_1 + X_2 > 1\}$.

14. Define the random variable X by

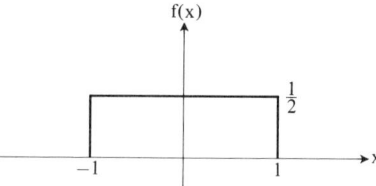

a) Compute and plot $F_X(x)$.
b) Compute and plot $F_X(x|M)$, where $M = \{3, 7, 8\}$.
c) Compute and plot $f_X(x|M)$.

15. The random variable X is a normal (0, 1). Find and plot $F_X(x|B)$ and $f_X(x|B)$ if
a) $B = \{X < 0\}$,   b) $B = \{X > 1\}$,   c) $B = \{X = 0\}$.

Fig. 7.23

16. The random variable X is uniformly distributed as shown in Fig. 7.23. Find the conditional density function $f(x|B)$ if
a) $B = \{X > 0\}$,   b) $B = \{X = 0\}$.

17. If X and Y are statistically independent random variables, what is the characteristic function of $Z = aX + bY$ in terms of the characteristic functions of X and Y?

156   RANDOM VARIABLES (*Continued*)

18. Computer assignment. In Problem 19 of Chapter 6 you were required to generate numbers $\{y_i\}$ derived from a uniform number generator.
   a) What is the value of the correlation coefficient ρ between the random variables Y_1 and Y_2? between Y_1 and Y_3?
   b) Generate 200 numbers $\{z_i\}$ as follows:

   $$z_1 = \sum_{i=1}^{12} x_i - 6.0, \qquad z_2 = \sum_{i=7}^{18} x_i - 6.0, \qquad z_3 = \sum_{i=13}^{24} x_i - 6.0, \qquad \ldots,$$

   where each x_i is from the uniform number generator. Find and plot an empirical cdf for these numbers. What is the correlation coefficient between Z_1 and Z_2? between Z_1 and Z_3?
   c) Generate 200 numbers $\{v_i\}$ as follows:

   $$v_1 = \sum_{i=1}^{12} x_i - 6.0, \qquad v_2 = \sum_{i=4}^{15} x_i - 6.0, \qquad v_3 = \sum_{i=7}^{18} x_i - 6.0, \qquad \ldots$$

   Find and plot an empirical cdf for these numbers. What is the correlation coefficient between V_1 and V_2? between V_1 and V_3?
   d) The empirical cdf should be approximately the same for each random variable Y, Z, and V in parts (a), (b), and (c) of this problem. That is, different values of correlation coefficient between successive random variables has no effect on the probability distribution.

   In order to discover what effect ρ has, plot the numbers, $y_1, y_2, \ldots, y_n, \ldots,$ y_{200} versus the index n. Do the same for $z_1, z_2, \ldots, z_{200}$ and for $v_1, v_2, \ldots, v_{200}$. Is there a relationship between the rate of variation from one number to the next and the correlation coefficient?

FURTHER READING

1. W. A. PORTER, *Modern Foundations of Systems Engineering*, Macmillan, New York, 1966.
2. P. L. MEYER, *Introductory Probability and Statistical Applications*, Addison-Wesley, Reading, Mass., 1965.
3. A. PAPOULIS, *Probability, Random Variables, and Stochastic Processes*, McGraw-Hill, New York, 1965.
4. J. M. WOZENCRAFT and I. M. JACOBS, *Principles of Communication Engineering*, Wiley, New York, 1965.

The text by Porter has an excellent discussion of function and composite function, which is the basis for our Sections 7.1, 7.2, and 7.3. Our approach to this material is based largely on the material in Papoulis. References 2, 3, and 4 each provide good discussions of the material in Sections 7.4 and 7.5.

8
LIMIT THEOREMS AND THE NORMAL DISTRIBUTION

8.1 LIMIT THEOREMS

We now wish to discuss some of the limit theorems that form the core of probability theory. These theorems are significant in their own right, but more importantly, the central limit theorem is a justification for the widespread use of the normal distribution.

Chebyshev's Inequality

The following theorem is known as Chebyshev's inequality.

Chebyshev's inequality. Let X be a random variable with $E(X) = 0$ and variance σ_X^2. Then for any positive number ϵ we have

$$P\{|X| \geq \epsilon\} \leq \frac{\sigma_X^2}{\epsilon^2} \qquad (8.1)$$

or, equivalently,

$$P\{|X| < \epsilon\} \geq 1 - \frac{\sigma_X^2}{\epsilon^2}. \qquad (8.2)$$

The proof of this powerful theorem is quite simple. Consider the mean-square value of X:

$$E(X^2) = \int_{-\infty}^{+\infty} x^2 f_X(x)\, dx. \qquad (8.3)$$

Now refer to Fig. 8.1 which shows a typical pdf with the values ϵ and $-\epsilon$

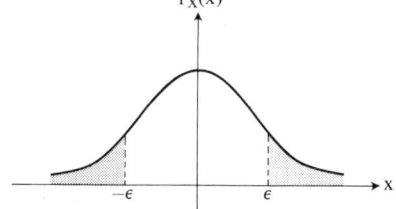

Fig. 8.1

marked on the graph. The area of the shaded portion is the probability $P\{|X| \geq \epsilon\}$. Since the integrand of Eq. (8.3) is positive we can write

$$E(X^2) \geq \int_{|x| \geq \epsilon} x^2 f_X(x)\, dx.$$

That is, the mean-square value of X is

$$E(X^2) = \int_{-\infty}^{-\epsilon} x^2 f_X(x)\, dx + \int_{-\epsilon}^{+\epsilon} x^2 f_X(x)\, dx + \int_{+\epsilon}^{\infty} x^2 f_X(x)\, dx$$

and we are simply leaving out the middle term. Now if x^2 is replaced by its smallest value ϵ^2, this inequality becomes even stronger:

$$E(X^2) \geq \epsilon^2 \int_{|x| \geq \epsilon} f_X(x)\, dx = \epsilon^2 P\{|X| \geq \epsilon\}.$$

Since $m_X = 0$, Eq. (8.1) follows immediately.

Notes. a) This is an extremely useful and powerful theorem, for look what it says! If you know only the mean and variance of the random variable Y, then you can make a statement about the probable value of Y. Simply subtract the mean from Y to obtain $X = Y - m_Y$ for use in either (8.1) or (8.2).

b) Don't infer that this inequality is good only for the gaussian distribution, as you might from looking at Fig. 8.1. This inequality applies to any random variable, continuous or discrete.

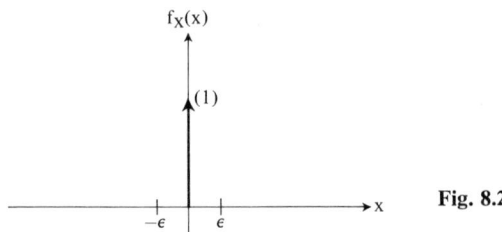

Fig. 8.2

c) Consider the degenerate case $\sigma_X^2 = 0$. Then Eq. (8.2) implies that $P\{|X| < \epsilon\} = 1$, since a probability can never be greater than one. This implies that for any experimental outcome the random variable X will assume its mean value, and hence the "width" of the pdf is zero. All the probability is concentrated at the mean, as in Fig. 8.2.

Examples

1. Let X be a random variable with mean $m_X = 0$ and variance $\sigma_X^2 = 2$. Find the largest probability that $|X| \geq 2$. From Eq. (8.1) this is

$$P\{|X| \geq 2\} \leq \tfrac{2}{4} = \tfrac{1}{2}.$$

2. Let Y be a random variable with mean $m_Y = 2$ and mean-square value $E(Y^2) = 5$. Find the smallest probability that $-1 \leq Y \leq 5$.

Subtract the mean from Y to obtain

$$X = Y - m_Y.$$

Then the events $\{-1 < Y < 5\}$ and $\{-3 < X < 3\}$ are equivalent:

$$\{-1 < Y < 5\} \sim \{|X| < 3\}. \tag{8.4}$$

Applying Chebyshev's inequality (Eq. 8.2), we have

$$P\{|X| < 3\} \geq 1 - (\sigma_X^2/9) = 1 - \tfrac{1}{9} = \tfrac{8}{9}$$

or, from Eq. (8.4),

$$P\{-1 < Y < 5\} \geq \tfrac{8}{9}.$$

Note. These results may be interpreted (loosely) as follows. Suppose the experiment is performed a large number of times. Then the value of X in Example 1 will be outside the interval $(-2, +2)$ somewhat less than half of the time.

If we knew the distribution of X (rather than just the first two moments), we could determine $P\{|X| \geq 2\}$ exactly. For example, if X is gaussian $N(0, 2)$, then $P\{|X| \geq 2\} = 0.3174$. If X is binary with zero mean and variance $\sigma_X^2 = 2$, then $P\{|X| \geq 2\} = 0$.

The Weak Law of Large Numbers

So far, in our discussion of probability, we have shown little or no direct relation between our definitions (for probability, etc.) and the real world. We have demonstrated that our models are useful in solving practical problems, but this is a result of coincidence (albeit planned) rather than a result of the inherent nature of our model. One direct link between the real world and our mathematical model is provided by the law of large numbers. [It is reassuring to know that such links do exist, for without them we could never be sure that our model was valid (except by the absence of experimental contradiction).]

For example, suppose a die is thrown a large number of times. One intuitively feels that the face with one dot on it should turn up roughly $\tfrac{1}{6}$ of the time. That is, if n is the number of times a 1 is rolled and N is the total number of times the die is thrown, then the ratio n/N should be close to $\tfrac{1}{6}$ for large N, and furthermore this ratio should get closer to $\tfrac{1}{6}$ as N increases. This is precisely the statement of the *law of large numbers*.

The *weak* law of large numbers follows directly from Chebyshev's inequality. Define the random variable Z as

$$Z = \frac{1}{N}\sum_{i=1}^{N} X_i, \tag{8.5}$$

where the $\{X_i\}$ are identically distributed random variables with mean $\bar{X}$ and variance σ_X^2. With reference to our die-throwing experiment, define the random variable X as

$$X = \begin{cases} 1 & \text{if a 1 is rolled,} \\ 0 & \text{otherwise.} \end{cases}$$

Then

$$n = \sum_{i=1}^{N} X_i \quad \text{and} \quad Z = \frac{n}{N}.$$

The mean of Z is

$$E(Z) = E\left(\frac{1}{N}\sum_{i=1}^{N} X_i\right) = \frac{1}{N}\sum_{i=1}^{N} E(X_i) = \bar{X}$$

and the variance is

$$\sigma_Z^2 = \frac{\sigma_X^2}{N}.$$

[We derive the variance of Z as follows: First we recognize that the variance of the sum of statistically independent random variables is the sum of the variances. That is,

$$\text{var}\left(\sum_{i=1}^{N} X_i\right) = \sum_{i=1}^{N} \sigma_X^2,$$

where var ( ) is the variance of ( ). Next, we have that the variance of a constant times a random variable is

$$\text{var}(aX) = a^2 \text{var}(X).$$

Therefore, we can write

$$\sigma_Z^2 = \frac{1}{N^2}\sum_{i=1}^{N} \sigma_X^2 = \frac{N\sigma_X^2}{N^2} = \frac{\sigma_X^2}{N},$$

as stated above.]

Now we wish to use Chebyshev's inequality, Eq. (8.1), on the random variable Z, so we define Y as

$$Y = Z - \bar{Z} = Z - \bar{X}.$$

The mean of Y is

$$E(Y) = E(Z - \bar{X}) = 0$$

and the variance is

$$\sigma_Y^2 = \sigma_Z^2 = \frac{\sigma_X^2}{N}.$$

Therefore Eq. (8.1) becomes

$$P\{|Y| \geq \epsilon\} \leq \frac{\sigma_Y^2}{\epsilon^2}$$

or

$$P\{|Z - \bar{X}| \geq \epsilon\} \leq \frac{\sigma_X^2}{N\epsilon^2}. \tag{8.6}$$

Equation (8.6) is called the *weak law of large numbers*.

Notes. a) The weak law of large numbers is a statement about the probable value of the sum of N identically distributed random variables. Specifically, if this sum is divided by N, then Eq. (8.6) states that the random variable Z is very likely to be near the mean value of X if N is large.

b) This law provides justification for our intuitive feeling that if an experiment is performed a large number of times, the relative frequency of a particular outcome should be near the probability of that outcome. Our example concerning the die throwing is continued below to illustrate this point.

c) The *weak* law of large numbers is opposed to the *strong* law of large numbers due to Borel. The weak law is concerned with statements about the probability of events such as $\{|Z - \bar{X}| < \epsilon\}$. Proof of these statements follows from Chebyshev's inequality. In contrast, the strong law allows one to make statements such as

$$P\{Z \to \bar{X}\} = 1 \quad \text{as} \quad N \to \infty,$$

and is simply a stronger statement of the weak law. The proof of the strong law is not easy, and is almost never given in elementary probability texts.

Now let us continue our example of the die-throwing experiment. With n/N defined as above, how many times must the die be thrown for the probability that n/N will be between 0 and $\frac{1}{3}$ to be greater than 0.5? That is, find N so that

$$P\{0 < (n/N) < \tfrac{1}{3}\} \geq 0.5.$$

With $Z = n/N$, the mean and variance of Z are

$$\bar{Z} = \tfrac{1}{6} \quad \text{and} \quad \sigma_Z^2 = \frac{\sigma_X^2}{N} = \frac{\tfrac{5}{36}}{N},$$

and ϵ is $\tfrac{1}{6}$. The problem can be restated as follows: Find N so that

$$P\{|Z - \bar{Z}| < \tfrac{1}{6}\} \geq 0.5.$$

Applying Chebyshev's inequality in the form of Eq. (8.2), we have

$$P\{|Z - \bar{Z}| < \tfrac{1}{6}\} \geq 1 - \frac{\tfrac{5}{36}}{N(\tfrac{1}{36})},$$

and if we set

$$(1 - 5/N) \geq 0.5,$$

we are assured that $P\{|Z - \bar{Z}| < \tfrac{1}{6}\} \geq 0.5$. Manipulating this inequality, we finally obtain

$$N \geq 10.$$

Note. This is a conservative estimate, for we used only the mean and variance of Z in arriving at this bound on N. If we had used more statistical information (for instance, the distribution of Z), we would have found a more accurate estimate.

As a further check on the validity of the law of large numbers, suppose the probability bound is changed from 0.5 to 0.8. That is, if we require $P\{0 < n/N < \frac{1}{3}\} \geq 0.8$, then the minimum N should be increased. Using this, our inequality becomes
$$(1 - 5/N) \geq 0.8,$$
so that $N \geq 25$.

The Central Limit Theorem

When discussing the binomial distribution, we were concerned with the sum of N independent random variables $X_1 + X_2 + \cdots + X_m$. The variance of this sum is $\sigma_{X_1}^2 + \sigma_{X_2}^2 + \sigma_{X_N}^2$ and is equal to $N\sigma_X^2$ if all the N random variables have the same variance. In our discussion of the law of large numbers above, we divided this sum by N, and we found that the variance of
$$Z = \frac{1}{N} \sum_{i=1}^{N} X_i$$
approaches zero as $N \to \infty$.

We now wish to discuss the behavior of the sum $\sum_{i=1}^{N} X_i$ as $N \to \infty$, particularly the tendency of this sum to approach the gaussian distribution. For this discussion the variance should become neither infinite nor zero, and this is accomplished by division by $\sqrt{N}$. Therefore let us define the random variable Y by
$$Y = \frac{1}{\sqrt{N}} \sum_{i=1}^{N} X_i. \tag{8.7}$$

If the variance of each X_i is the same, then the variance of Y is equal to σ_X^2 for any N.

> **Central limit theorem.** Suppose the X_i of Eq. (8.7) are statistically independent random variables with zero mean and finite variance, and that all have identical distribution. Then for any t,
> $$\lim_{N \to \infty} F_Y(t) = \int_{-\infty}^{t} \frac{1}{\sqrt{2\pi}\,\sigma_X} e^{-\lambda^2/2\sigma_X^2}\, d\lambda. \tag{8.8}$$

Notes. a) The theorem states that the *area* under the pdf f_Y is equal to the *area* under an appropriate gaussian pdf. Note specifically that the theorem does *not* say that the pdf f_Y is equal to a gaussian pdf in the limit. As a counterexample, suppose each X_i has binary distribution. Then the pdf of Y consists of impulses and can never approach the smooth gaussian pdf.

b) Note the assumptions we made in the statement of the theorem, namely, that all X_i must be identically distributed and statistically independent. There are several degrees of complexity for the central limit theorem, depending on the assumptions made. For example, the distribution of all X_i need

8.1 LIMIT THEOREMS

not be identical. We have chosen the above form to simplify the following discussion.

The central limit theorem is really a property of convolution, where we are convolving a large number of functions. Consider the sum of two statistically independent random variables, say $X = X_1 + X_2$. Then the pdf f_X is just the convolution of f_{X_1} with f_{X_2}: $f_X = f_{X_1} * f_{X_2}$. This is shown by finding the characteristic function for X, as follows:

$$\phi_X(\nu) = E(e^{j\nu X}) = E(e^{j\nu(X_1+X_2)}) = E(e^{j\nu X_1})E(e^{j\nu X_2}) = \phi_{X_1}(\nu)\phi_{X_2}(\nu).$$

Since the characteristic function of X is the product of the characteristic functions of X_1 and X_2, the pdf f_X is the convolution of the two pdf's f_{X_1} and f_{X_2}. This follows from the convolution property of Fourier transforms, Eq. (3.17). Now the random variable Y in the theorem is related to the sum of N statistically independent random variables. We can expect the pdf f_Y to somehow be related to the convolution of the N pdf's f_{X_i}, and the characteristic function ϕ_Y should be related to the product of the N characteristic functions ϕ_{X_i}.

To proceed, let us derive the relationship between ϕ_Y and ϕ_X:

$$\phi_Y(\nu) = E(e^{j\nu Y}) = E\left[\exp\left(j\nu \frac{1}{\sqrt{N}} \sum_{i=1}^{N} X_i\right)\right]$$

$$= E\left[\prod_{i=1}^{N} \exp\left(j\nu \frac{X_i}{\sqrt{N}}\right)\right]$$

$$= \prod_{i=1}^{N} E\left[\exp\left(j\nu \frac{X_i}{\sqrt{N}}\right)\right] = \left[\phi_X\left(\frac{\nu}{\sqrt{N}}\right)\right]^N. \quad (8.9)$$

The remainder of our discussion is quite simple, though the details are complicated. We wish to show that as $N \to \infty$ the characteristic function ϕ_Y is of the form of the gaussian characteristic function.

The Maclaurin series for $e^{j\nu x}$ is

$$e^{j\nu x} = 1 + j\nu x + \frac{(j\nu)^2}{2!} x^2 + \frac{(j\nu)^3}{3!} x^3 + \cdots$$

Assume that all moments of X are finite. Then

$$E(e^{j\nu X}) = 1 + j\nu \bar{X} + \frac{(j\nu)^2}{2!} \overline{X^2} + \frac{(j\nu)^3}{3!} \overline{X^3} + \cdots = \phi_X(\nu),$$

but $\bar{X} = 0$, so that $\overline{X^2} = \sigma_X^2$. Therefore $\phi_X(\nu) = 1 - (\nu^2/2!)\sigma_X^2 +$ terms involving ν^3 and higher powers of ν. For simplicity, we will denote these terms involving ν^3 and higher powers of ν as $\nu^3 f(\nu)$. Then

$$\phi_X(\nu) = 1 - \frac{\nu^2}{2!} \sigma_X^2 + \nu^3 f(\nu). \quad (8.10)$$

164   LIMIT THEOREMS AND THE NORMAL DISTRIBUTION   8.2

Now let us return to the relation between ϕ_Y and ϕ_X, Eq. (8.9). Taking logarithms and substituting Eq. (8.10), we have

$$\ln \phi_Y(v) = N \ln \phi_X\left(\frac{v}{\sqrt{N}}\right) = N \ln \left[1 - \frac{v^2}{N}\frac{\sigma_X^2}{2} + \left(\frac{v}{\sqrt{N}}\right)^3 f\left(\frac{v}{\sqrt{N}}\right)\right]. \quad (8.11)$$

Let us digress to point out that an expansion for $\ln(1 + \alpha)$ is

$$\ln(1 + \alpha) = \alpha - \frac{\alpha^2}{2} + \frac{\alpha^3}{3} - \frac{\alpha^4}{4} + \cdots, \quad |\alpha| < 1.$$

Using this we can write Eq. (8.11) as

$$N \ln \phi_X\left(\frac{v}{\sqrt{N}}\right)$$

$$= N\left\{-\frac{v^2}{N}\frac{\sigma_X^2}{2} + \left(\frac{v}{\sqrt{N}}\right)^3 f\left(\frac{v}{\sqrt{N}}\right) + \text{terms of } N^{-2}, N^{-3/2}, N^{-3}, \ldots\right\}. \quad (8.12)$$

The condition $|\alpha| < 1$ is satisfied by choosing N large enough. After all, we are going to use this expression in the limit as $N \to \infty$.

To complete our discussion, we take the limit as $N \to \infty$ in Eq. (8.12):

$$\lim_{N \to \infty} \ln \phi_Y(v) = \lim_{N \to \infty} N \ln \phi_X\left(\frac{v}{\sqrt{N}}\right) = -\frac{v^2 \sigma_X^2}{2}$$

or, raising both sides to the exponential power,

$$\lim_{N \to \infty} \phi_Y(v) = e^{-v^2 \sigma_X^2/2}, \quad (8.13)$$

which we recognize as the characteristic function of a gaussian random variable. (See Eq. 6.44.)

Note. The above derivation is not very precise. We took several questionable steps without justification, and although we could justify these steps, we must not call this derivation a proof. The above should be regarded as an outline or explanation. Almost any elementary probability text presents the same argument. The particular argument above is due to Wozencraft and Jacobs [1]. For a readable proof, see Cramér [5], pages 213 to 218.

8.2 THE GAUSSIAN RANDOM VARIABLE

In Section 6.5 we discussed the one-dimensional gaussian random variable. Here we will spend some time on the important two-dimensional gaussian random variable and then extend these concepts to random variables of arbitrary dimensions.

Let X_1 and X_2 be two random variables defined on the same sample space, and suppose that X_1 and X_2 are each (marginally) normal with zero mean and unit variance. Then the two-dimensional pdf is of the form

$$f_{X_1X_2}(x_1, x_2) = \frac{1}{2\pi\sqrt{1-\rho^2}} \exp\left[-\frac{x_1^2 - 2\rho x_1 x_2 + x_2^2}{2(1-\rho^2)}\right]. \quad (8.14)$$

Notes. a) The factor ρ is the correlation coefficient of Eq. (7.44).

b) Equation (8.14) is only for the special case of zero mean and unit variance for each random variable. More general conditions will be explored below.

The correlation coefficient ρ deserves some attention. Figure 8.3 illustrates the two-dimensional pdf of Eq. (8.14) for different values of ρ.

Notes. a) Regardless of the value of ρ (so long as $-1 \leq \rho \leq 1$), the marginal density of each random variable X_1 and X_2 is still $N(0, 1)$.

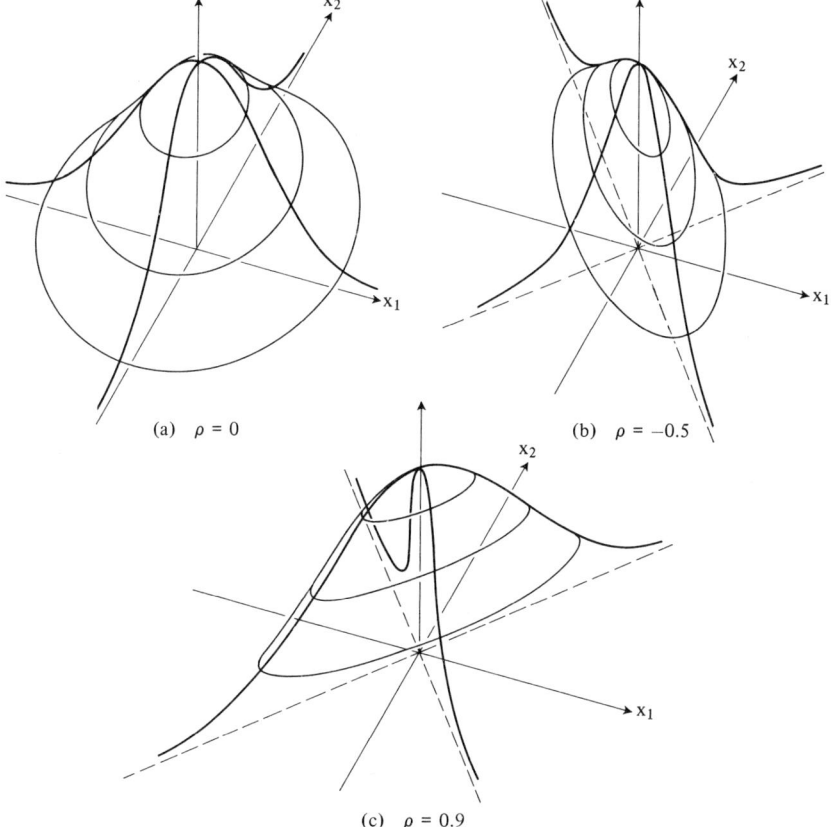

Fig. 8.3 Examples of the two-dimensional gaussian density function. (Adapted from Wozencraft and Jacobs, *Principles of Communication Engineering*, Wiley, New York, 1965. With permission.)

b) Recall that the term "density" implies that we must integrate to find probability. Thus the probability that the values of X_1 and X_2 will lie in some region of the $x_1 x_2$-plane is the volume under the curve $f_{X_1 X_2}$ over that region. The entire volume is one.

The Marginal Density

We will derive the marginal density for, say, X_1 from the joint pdf. This will illustrate our statement that $X_1 \sim N(0, 1)$ and also will be of use in the future. According to Eq. (7.40),

$$f_{X_1}(x_1) = \int_{-\infty}^{+\infty} f_{X_1 X_2}(x_1, x_2) \, dx_2$$

$$= \int_{-\infty}^{+\infty} \frac{1}{2\pi\sqrt{1 - \rho^2}} \exp\left[-\frac{x_1^2 - 2\rho x_1 x_2 + x_2^2}{2(1 - \rho^2)}\right] dx_2. \quad (8.15)$$

Now change the variable of integration. Let

$$\lambda = \frac{x_2 - \rho x_1}{\sqrt{1 - \rho^2}}. \quad (8.16)$$

The exponent in Eq. (8.15) can be expressed in terms of $(x_2 - \rho x_1)$ by noting that

$$(x_2 - \rho x_1)^2 + x_1^2(1 - \rho^2) = x_1^2 - 2\rho x_1 x_2 + x_2^2,$$

so that Eq. (8.15) becomes

$$f_{X_1}(x_1) = \frac{e^{-x_1^2/2}}{\sqrt{2\pi}} \int_{-\infty}^{+\infty} \frac{1}{\sqrt{2\pi(1 - \rho^2)}} \exp\left[-\frac{(x_2 - \rho x_1)^2}{2(1 - \rho^2)}\right] dx_2.$$

The substitution of λ from Eq. (8.16) now gives the desired result:

$$f_{X_1}(x_1) = \frac{e^{-x_1^2/2}}{\sqrt{2\pi}} \int_{-\infty}^{+\infty} \frac{1}{\sqrt{2\pi}} e^{-\lambda^2/2} \, d\lambda$$

$$= \frac{1}{\sqrt{2\pi}} e^{-x_1^2/2},$$

and this is recognized as the one-dimensional gaussian form with zero mean and unit variance.

The Conditional Density

The conditional pdf was defined in Section 7.5 (Definition 7) for general random variables. Here we amplify that discussion and apply the definition to gaussian random variables. Suppose we are allowed to observe one of the random variables, say X_2, and that the observed value of X_2 is v. This additional information should change the marginal pdf of X_1. What we are saying is that

8.2 THE GAUSSIAN RANDOM VARIABLE

the conditional pdf $f_{X_1}(x_1|X_2 = v)$ should be different from the unconditional (marginal) pdf $f_{X_1}(x_1)$. This statement may not be obvious, so we will present an intuitive argument based on Fig. 8.3(b) before proceeding to derive the conditional pdf.

Suppose $X_2 = -0.5$. Now imagine that the two-dimensional pdf is cut vertically along a line parallel to the x_1-axis and passing through $x_2 = -0.5$. Figure 8.3(b) is redrawn in Fig. 8.4 to show this silhouette.

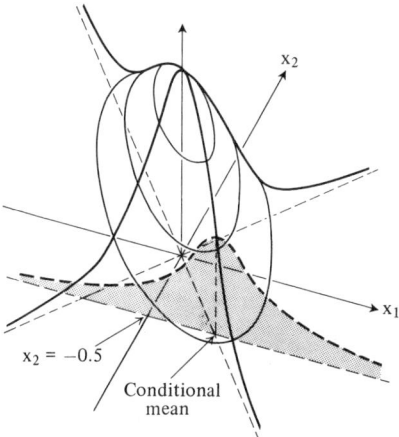

Fig. 8.4

The area under the silhouette is not equal to one. If it were, this would be the conditional pdf $f_{X_1}(x_1|X_2 = -0.5)$. But the area is easily adjusted to one by a normalizing constant. Note that the mean is not zero.

From Definition 7 of Chapter 7, the conditional pdf of X_1 given that $X_2 = v$ is

$$f_{X_1}(x_1|X_2 = v) = \frac{f_{X_1X_2}(x_1, v)}{f_{X_2}(v)}. \tag{8.17}$$

Notes. a) The denominator of Eq. (8.17) is a number. This is the height of f_{X_2} evaluated at the particular value v. This number is sometimes called the *normalizing factor*, for it serves the purpose of making the area under $f_{X_1}(x_1|X_2 = v)$ equal to one.

b) The numerator of Eq. (8.17) forms the silhouette (such as in Fig. 8.4) as x_1 is varied and x_2 is held constant at the value of v. The area under this silhouette is not one, but division by $f_{X_2}(v)$ makes it so.

c) Do not conclude that Eq. (8.17) applies only to the gaussian random variable. It is necessary only that $f_{X_2}(v)$ be continuous at $X_2 = v$.

Now we shall derive the conditional pdf of the gaussian random variable. Suppose $f_{X_1 X_2}$ is given by Eq. (8.14). Then the conditional pdf is

$$f_{X_1}(x_1 | X_2 = v) = \frac{f_{X_1 X_2}(x_1, v)}{f_{X_2}(v)}$$

$$= \frac{\dfrac{1}{2\pi\sqrt{1-\rho^2}} \exp\left[-\dfrac{x_1^2 - 2\rho x_1 v + v^2}{2(1-\rho^2)}\right]}{\dfrac{1}{\sqrt{2\pi}} \exp\left[-\dfrac{v^2}{2}\right]}$$

$$= \frac{1}{\sqrt{2\pi(1-\rho^2)}} \exp\left[-\frac{x_1^2 - 2x_1 v + v^2}{2(1-\rho^2)} + \frac{v^2}{2}\right]$$

$$= \frac{1}{\sqrt{2\pi(1-\rho^2)}} \exp\left[-\frac{(x_1 - \rho v)^2}{2(1-\rho^2)}\right]. \qquad (8.18)$$

Note. This is in the form of Eq. (6.31), where the mean is ρv and the variance is $(1 - \rho^2)$. Thus the conditional gaussian pdf has the mean shifted from 0 to ρv and the variance reduced from 1 to $(1 - \rho^2)$.

The Two-Dimensional Gaussian pdf with Arbitrary Mean and Variance

First, suppose the means of X_1 and X_2 are still zero and the variances are equal but arbitrary. Then

$$f_{X_1 X_2}(x_1, x_2) = \frac{1}{2\pi\sigma^2\sqrt{1-\rho^2}} \exp\left[-\frac{x_1^2 - 2\rho x_1 x_2 + x_2^2}{2\sigma^2(1-\rho^2)}\right]. \qquad (8.19)$$

Next, suppose the variances are unequal. Then the joint pdf is given by

$$f_{X_1 X_2}(x_1, x_2) = \frac{1}{2\pi\sigma_1\sigma_2\sqrt{1-\rho^2}} \exp\left\{-\frac{1}{2(1-\rho^2)}\left[\frac{x_1^2}{\sigma_1^2} - \frac{2\rho x_1 x_2}{\sigma_1\sigma_2} + \frac{x_2^2}{\sigma_2^2}\right]\right\}. \qquad (8.20)$$

And finally, if the means are m_{X_1} and m_{X_2}, we replace x_1 by $(x_1 - m_{X_1})$ and x_2 by $(x_2 - m_{X_2})$ in Eq. (8.20) to yield the general form. Thus we obtain

$$f_{X_1 X_2}(x_1, x_2) = \frac{1}{2\pi\sigma_1\sigma_2\sqrt{1-\rho^2}} \exp\left\{-\frac{1}{2(1-\rho^2)}\left[\frac{(x_1 - m_{X_1})^2}{\sigma_1^2}\right.\right.$$
$$\left.\left. - \frac{2\rho}{\sigma_1\sigma_2}(x_1 - m_{X_1})(x_2 - m_{X_2}) + \frac{(x_2 - m_{X_2})^2}{\sigma_2^2}\right]\right\}. \qquad (8.21)$$

Properties of the Gaussian Random Variable

1. The gaussian distribution depends *only* on the first- and second-order moments of the random variables. that is, only on the means, variances, and covariances.
2. If X_1 and X_2 are jointly gaussian, they are individually gaussian. Both the one-dimensional marginal and the conditional distributions are gaussian.
3. If two gaussian random variables are uncorrelated, they are statistically independent. (The converse is true for any distribution.) This means that if $E(X_1 X_2) = E(X_1)E(X_2)$, then the gaussian random variables X_1 and X_2 are statistically independent. To see this, consider any of the joint functions in Eqs. (8.14), (8.19), (8.20), and (8.21). With $\rho = 0$, we have

$$f_{X_1 X_2}(x_1, x_2) = f_{X_1}(x_1) f_{X_2}(x_2),$$

which is the condition for statistical independence.

4. Linear transformations on gaussian random variables result in gaussian random variables. For example, consider

$$Y = aX + b.$$

If X is gaussian, then so is Y. For another example, consider

$$Y = aX_1 + bX_2.$$

If X_1 and X_2 are jointly gaussian, then Y is a one-dimensional gaussian random variable.

Property 4 is extremely useful in the study of linear systems. If the statistics of the input are gaussian, then we are assured that the output is also gaussian.

The Multivariate Gaussian Random Variable

Note that the formulas grew progressively complex as the joint normal distribution was generalized from Eq. (8.19) through Eq. (8.21). To handle third-, fourth-, and higher-dimensional gaussian random variables, we will need matrix notation. A short appendix is included in this chapter for those readers who are unfamiliar with matrices.

Given an n-dimensional random vector $\mathbf{X} = (X_1, X_2, \ldots, X_n)$ consisting of n statistically independent random variables with zero mean, their joint pdf is the product

$$f_{\mathbf{X}}(\mathbf{x}) = \prod_{i=1}^{n} f_{X_i}(x_i) = \prod_{i=1}^{n} \frac{1}{\sqrt{2\pi\sigma_i^2}} e^{-x_i^2/2\sigma_i^2}.$$

In matrix notation this becomes

$$f_{\mathbf{X}}(\mathbf{x}) = \frac{1}{(2\pi)^{n/2}|\Lambda_X|^{1/2}} \exp\left(-\tfrac{1}{2}\mathbf{x}\Lambda_X^{-1}\mathbf{x}^T\right), \qquad (8.22)$$

where $\mathbf{x}$ is the n-dimensional row vector

$$\mathbf{x} = (x_1, x_2, \ldots, x_n),$$

$\mathbf{x}^T$ is the transpose, and Λ_X is the matrix

$$\Lambda_X = \begin{bmatrix} \sigma_1^2 & 0 & \cdots & 0 \\ 0 & \sigma_2^2 & \cdots & 0 \\ \vdots & \vdots & & \vdots \\ 0 & 0 & \cdots & \sigma_n^2 \end{bmatrix}.$$

Notes. a) The quadratic form $Q = \mathbf{x}\Lambda_X^{-1}\mathbf{x}^T$ in Eq. (8.22) is just a number. That is, for a given $\mathbf{x}$ and Λ_X, Q is a number.

b) Since the random variables $(X_1, X_2, \ldots, X_n)$ are statistically independent, Λ_X is a diagonal matrix. The converse is also true: If Λ_X is a diagonal matrix, the random variables are independent.

The matrix Λ_X is called the *covariance matrix* and consists of the terms $\lambda_{ij} = E[(X_i - \bar{X}_i)(X_j - \bar{X}_j)] = \overline{X_i X_j} - \bar{X}_i \bar{X}_j$. The term λ_{ij} is called the *covariance* when $i \ne j$. Note that when $i = j$, λ_{ij} is the variance σ_i^2.

In general, the covariance matrix Λ_X is not diagonal. That is, X_i and X_j are correlated and statistically dependent, so that the off-diagonal terms are nonzero. In this case the multidimensional gaussian pdf is still in the form of Eq. (8.22).

For random variables with nonzero mean, replace the vector $\mathbf{x}$ in Eq. (8.22) by $(\mathbf{x} - \mathbf{m}_X)$. Thus we get

$$f_{\mathbf{X}}(\mathbf{x}) = \frac{1}{(2\pi)^{n/2}|\Lambda_X|^{1/2}} \exp\left[-\tfrac{1}{2}(\mathbf{x} - \mathbf{m}_X)\Lambda_X^{-1}(\mathbf{x} - \mathbf{m}_X)^T\right]. \qquad (8.23)$$

Note. For a given vector $\mathbf{x}$, the quadratic form $Q = (\mathbf{x} - \mathbf{m}_X)\Lambda_X^{-1}(\mathbf{x} - \mathbf{m}_X)^T$ is just a number.

Equation (8.23) is the general form for the multidimensional gaussian pdf. As an example, consider the general two-dimensional pdf of Eq. (8.21). The notation there can be simplified by defining Λ_X as

$$\Lambda_X = \begin{bmatrix} \sigma_1^2 & \rho\sigma_1\sigma_2 \\ \rho\sigma_1\sigma_2 & \sigma_2^2 \end{bmatrix},$$

so that $|\Lambda_X| = \sigma_1^2\sigma_2^2(1 - \rho^2)$ and the inverse is

$$\Lambda_X^{-1} = \frac{1}{|\Lambda_X|}\begin{bmatrix} \sigma_2^2 & -\rho\sigma_1\sigma_2 \\ -\rho\sigma_1\sigma_2 & \sigma_1^2 \end{bmatrix} = \frac{1}{1-\rho^2}\begin{bmatrix} 1/\sigma_1^2 & -\rho/\sigma_1\sigma_2 \\ -\rho/\sigma_1\sigma_2 & 1/\sigma_2^2 \end{bmatrix}.$$

After carrying out the indicated matrix multiplication in Eq. (8.23), we see that the expansion yields Eq. (8.21).

PROBLEMS

1. Suppose $E(X) = 2$ and $E(X^2) = 8$.
   a) What is the variance of X?
   b) Find the smallest probability that $-1 < X < 5$.

2. Two independent random variables X_1 and X_2 are added together. The distribution of each X is uniform $(-1, 1)$. Find *and plot* the pdf of the sum.

3. Repeat Problem 2 for three random variables X_1, X_2, X_3.

4. Repeat Problem 3 for the sum divided by the square root of 3.

5. Repeat Problem 2 for the sum of four random variables divided by the square root of 4.

6. Show that the sum of N random variables, each with gaussian distribution and arbitrary mean and variance, is also gaussian.

7. Let X_1 and X_2 be independent random variables, each with uniform distribution $(0, 1)$. Find the distribution of $Y = X_1 + 2X_2$.

8. Suppose a coin is tossed N times and the number of times heads are obtained is given by n_H. Find the number N of tosses necessary to have a probability greater than 0.9 that $0.4 < n_H/N < 0.6$.

9. Let X be a gaussian random variable with zero mean and unit variance. Find $P\{-1 < X < 2\}$.

10. Let X_1 and X_2 be independent gaussian random variables with zero mean and unit variance. For each part of Fig. 8.5, find the probability that X_1 and X_2 lie in the shaded region.

11. Let $X = (X_1, X_2)$ be a gaussian random variable with zero mean and covariance matrix

$$\Lambda_X = \begin{bmatrix} 2 & 1 \\ 1 & 4 \end{bmatrix}.$$

   a) Write an expression for the two-dimensional pdf.
   b) Determine the conditional density function $f_X(x_1|X_2 = 2)$.

12. Repeat Problem 11 given that $\overline{X}_1 = 1$, $\overline{X}_2 = -3$.

13. Determine the probability that the sum of the random variables of Problem 12 is greater than zero.

172   LIMIT THEOREMS AND THE NORMAL DISTRIBUTION

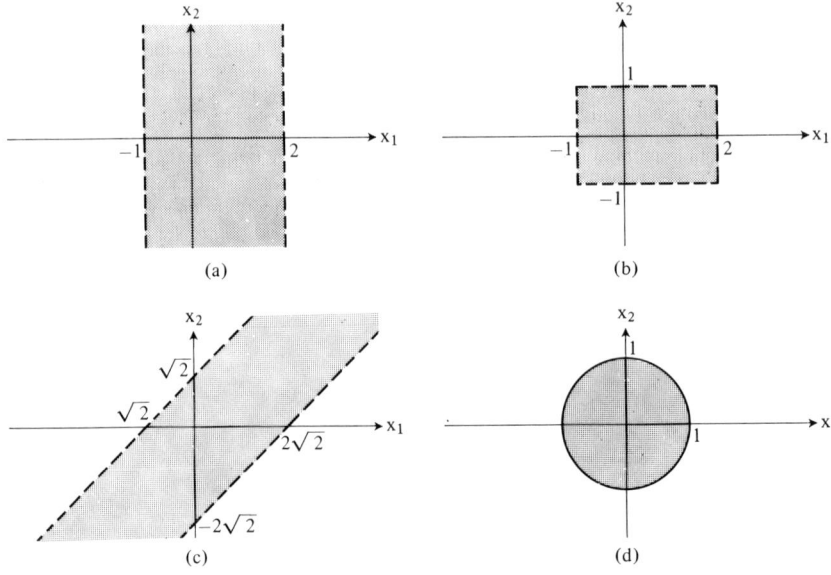

Fig. 8.5

14. If a coin is tossed 10,000 times, estimate $P\{4900 < n < 5100\}$, where n is the number of heads.
   a) Use Chebyshev's inequality to obtain a bound on this probability.
   b) Use the central limit theorem to estimate this probability.

FURTHER READING

1. J. M. WOZENCRAFT and I. M. JACOBS, *Principles of Communication Engineering*, Wiley, New York, 1965.
2. P. L. MEYER, *Introductory Probability and Statistical Applications*, Addison-Wesley, Reading, Mass., 1965.
3. F. AYRES, JR., *Outline of Theory and Problems of Matrices*, Schaum, New York, 1962.
4. F. A. GRAYBILL, *Linear Statistical Models*, McGraw-Hill, New York, 1961.
5. H. CRAMÉR, *Mathematical Methods of Statistics*, Princeton University Press, Princeton, 1946.

References 1 and 2 provide a good discussion of limit theorems, and 2 has an especially good discussion of the gaussian random variable. Both 3 and 4 provide good material on matrices. Reference 5 provides a readable proof of the central limit theorem.

APPENDIX: MATRICES

Definition

An *n*-by-*m* ($n \times m$) matrix is a rectangular array of numbers with n rows and m columns. For example, the $n \times m$ matrix **A** composed of elements a_{ij} is given by

$$\mathbf{A} = \begin{bmatrix} a_{11} & a_{12} & \cdots & a_{1m} \\ a_{21} & a_{22} & \cdots & a_{2m} \\ \vdots & \vdots & & \vdots \\ a_{n1} & a_{n2} & \cdots & a_{nm} \end{bmatrix},$$

where the *i*-subscript indicates the row of the element and the *j*-subscript indicates its column.

Notes. a) We shall use boldface capital letters to denote matrices and the corresponding lightface lower-case letters to denote the elements.

b) The array must be rectangular. The arrangements B and C, where

$$B = \begin{bmatrix} 1 & 4 & 10 \\ 6 & 2 & \\ 3 & & \end{bmatrix} \quad \text{and} \quad C = \begin{bmatrix} & & 10 & \\ & 2 & 3 & \\ 6 & 4 & & 10 \end{bmatrix},$$

are not matrices.

A matrix with n rows and n columns is called a *square matrix*. A $1 \times n$ matrix is called a *row vector*. An $n \times 1$ matrix is called a *column vector*. The following are examples of these three special matrices:

$$\text{square matrix:} \quad \mathbf{A} = \begin{bmatrix} a_{11} & a_{12} \\ a_{21} & a_{22} \end{bmatrix},$$

$$\text{row vector:} \quad \mathbf{B} = [b_1 \; b_2 \; b_3 \; \cdots \; b_n],$$

$$\text{column vector:} \quad \mathbf{C} = \begin{bmatrix} c_1 \\ c_2 \\ c_3 \\ \vdots \\ c_n \end{bmatrix}.$$

Note that in the row vector we need to indicate only the column of the elements, since all elements are in the same row. Similarly, we need to indicate only the row of the elements in the column vector.

Addition

Two matrices **A** and **B** can be added only if they both have the same number of rows and columns. The new matrix **C** = **A** + **B** is formed by adding corresponding elements. For example,

$$\begin{bmatrix} 1 & 6 \\ \frac{2}{3} & 0 \\ 3 & 1 \end{bmatrix} + \begin{bmatrix} \frac{1}{2} & 10 \\ 1 & 0 \\ \frac{1}{2} & 1 \end{bmatrix} = \begin{bmatrix} 1\frac{1}{2} & 16 \\ 1\frac{2}{3} & 0 \\ 3\frac{1}{2} & 2 \end{bmatrix}.$$

Scalar Multiplication

The product of a scalar (number) c and a matrix **A** is a new matrix **B** given by

$$\mathbf{B} = c\mathbf{A} = c \begin{bmatrix} a_{11} & a_{12} & \cdots & a_{1m} \\ a_{21} & a_{22} & \cdots & a_{2m} \end{bmatrix}$$

$$= \begin{bmatrix} ca_{11} & ca_{12} & \cdots & ca_{1m} \\ ca_{21} & ca_{22} & \cdots & ca_{2m} \end{bmatrix}.$$

That is, each element a_{ij} of the matrix **A** is multiplied by the scalar c.

Matrix Multiplication

The product of two matrices **A** and **B** is defined only if the number of columns in **A** is equal to the number of rows in **B**. If **A** is $n \times k$ and **B** is $k \times m$, then the product **C** = **AB** is $n \times m$:

$$\underset{(n \times k)}{\mathbf{A}} \underset{(k \times m)}{\mathbf{B}} = \underset{(n \times m)}{\mathbf{C}}.$$

Each element of **C** is found by the formula

$$c_{ij} = \sum_{l=1}^{k} a_{il} b_{lj}.$$

To obtain the ijth element in the product matrix **C**, multiply each element in the ith row of **A** by the corresponding element in the jth column of **B** and add these products. Consider the product of the two matrices **A** and **B** given by

$$\mathbf{A} = \begin{bmatrix} 2 & 3 \\ 1 & 3 \\ 6 & 2 \end{bmatrix} \quad \text{and} \quad \mathbf{B} = \begin{bmatrix} 1 & 3 \\ 2 & 1 \end{bmatrix}.$$

Since **A** is 3×2 and **B** is 2×2, the product **AB** is 3×2. (Note that we cannot form the product **BA**, since the number of columns in **B** is not equal to the number of rows in **A**.) The element c_{21} is found from the elements in the shaded portions of **A** and **B**:

$$c_{21} = a_{21}b_{11} + a_{22}b_{21} = 1 \cdot 1 + 3 \cdot 2 = 7.$$

The entire product matrix is shown below:

$$\mathbf{C} = \mathbf{AB} = \begin{bmatrix} 2+6 & 6+3 \\ 1+6 & 3+3 \\ 6+4 & 18+2 \end{bmatrix} = \begin{bmatrix} 8 & 9 \\ 7 & 6 \\ 10 & 20 \end{bmatrix}.$$

As another example, the product of a $1 \times n$ matrix and an $n \times 1$ matrix is a number (a 1×1 matrix). Considering the following:

$$[3 \quad 1 \quad 6] \begin{bmatrix} 1 \\ 2 \\ 1 \end{bmatrix} = 3 + 2 + 6 = 11.$$

Determinants

A matrix is not a number but an array of numbers. There is no such thing as the *value* of a matrix. On the other hand, a determinant is a number determined by a square matrix. The determinant of **A** is written

$$|\mathbf{A}| = \begin{vmatrix} a_{11} & a_{12} & \cdots & a_{1n} \\ a_{21} & a_{22} & \cdots & a_{2n} \\ \cdot & \cdot & & \cdot \\ \cdot & \cdot & & \cdot \\ \cdot & \cdot & & \cdot \\ a_{n1} & a_{n2} & \cdots & a_{nn} \end{vmatrix}.$$

The vertical bars on both the symbol **A** and the array of numbers are used to denote the determinant.

A determinant of order n is the determinant of an $n \times n$ matrix. We will indicate one method used to evaluate low-order determinants, and although this method is applicable to determinants of any order, the labor involved increases rapidly for high-order determinants. It should be noted that there are computational routines available for evaluating determinants that greatly reduce the labor involved with high-order determinants.

We will need the concepts of *minor* and *cofactor*. Let **A** be an $n \times n$ matrix as indicated above. When the ith row and jth column are removed from **A**, the determinant of the remaining $(n-1) \times (n-1)$ matrix is called the *minor* of a_{ij}. The signed minor found by multiplying $(-1)^{i+j}$ times the minor of a_{ij} is called the *cofactor* of a_{ij} and is denoted by A_{ij}.

Consider the 3 × 3 matrix **B** given by

$$\mathbf{B} = \begin{bmatrix} 1 & 3 & 2 \\ 6 & 1 & 2 \\ 3 & 4 & 1 \end{bmatrix}.$$

The minor of b_{12} is the determinant of the matrix formed from the unshaded elements,

$$\text{minor of } b_{12} = \begin{vmatrix} 6 & 2 \\ 3 & 1 \end{vmatrix},$$

and the cofactor of b_{12} is given by

$$B_{12} = (-1)^{1+2} \begin{vmatrix} 6 & 2 \\ 3 & 1 \end{vmatrix} = - \begin{vmatrix} 6 & 2 \\ 3 & 1 \end{vmatrix}.$$

The determinant of an $n \times n$ matrix can be found in terms of determinants of 2 × 2 matrices by use of cofactors. The determinant of a second-order matrix is given by

$$\begin{vmatrix} a_{11} & a_{12} \\ a_{21} & a_{22} \end{vmatrix} = a_{11}a_{22} - a_{12}a_{21}.$$

The determinant of a third-order matrix is given by

$$\begin{vmatrix} a_{11} & a_{12} & a_{13} \\ a_{21} & a_{22} & a_{23} \\ a_{31} & a_{32} & a_{33} \end{vmatrix} = a_{11}A_{11} + a_{12}A_{12} + a_{13}A_{13}$$

$$= a_{11} \begin{vmatrix} a_{22} & a_{23} \\ a_{32} & a_{33} \end{vmatrix} - a_{12} \begin{vmatrix} a_{21} & a_{23} \\ a_{31} & a_{33} \end{vmatrix} + a_{13} \begin{vmatrix} a_{21} & a_{22} \\ a_{31} & a_{33} \end{vmatrix}.$$

Note that the elements in the first row are multiplied by their cofactors and summed. Our use of the first row was arbitrary; we could have used any row or column. In general, the determinant of an $n \times n$ matrix is given by

$$|\mathbf{A}| = \sum_{i=1}^{n} a_{ij}\mathbf{A}_{ij} \quad \text{for fixed but arbitrary } j,$$

$$= \sum_{j=1}^{n} a_{ij}\mathbf{A}_{ij} \quad \text{for fixed but arbitrary } i.$$

The evaluation of a third-order determinant is reduced to the evaluation of three second-order determinants by this method. Likewise, the evaluation of a fourth-order determinant is reduced to the evaluation of four third-order determinants, each of which can be reduced to three second-order determinants. And so forth.

For example, the determinant of **B** is found by using the second column elements and their cofactors as follows:

$$|\mathbf{B}| = \begin{vmatrix} 1 & 3 & 2 \\ 6 & 1 & 2 \\ 3 & 4 & 1 \end{vmatrix} = -3\begin{vmatrix} 6 & 2 \\ 3 & 1 \end{vmatrix} + 1\begin{vmatrix} 1 & 2 \\ 3 & 1 \end{vmatrix} - 4\begin{vmatrix} 1 & 2 \\ 6 & 2 \end{vmatrix}$$
$$= 0 - 5 + 40 = 35.$$

The Transpose

The transpose of a matrix is formed by replacing each element a_{ij} by the element a_{ji}. The transpose of a row vector is a column vector and conversely. The notation $\mathbf{A}^T$ is used to denote the transpose of **A**. For example,

$$\mathbf{A} = \begin{bmatrix} 2 & 3 \\ 1 & 3 \\ 6 & 2 \end{bmatrix}, \quad \mathbf{A}^T = \begin{bmatrix} 2 & 1 & 6 \\ 3 & 3 & 2 \end{bmatrix},$$

and

$$\mathbf{B} = \begin{bmatrix} 1 & 3 \\ 2 & 1 \end{bmatrix}, \quad \mathbf{B}^T = \begin{bmatrix} 1 & 2 \\ 3 & 1 \end{bmatrix}.$$

It is evident that if **A** is $n \times m$, then $\mathbf{A}^T$ is $m \times n$. Although we can form the product **AB** for the above matrices, we cannot form the product $\mathbf{A}^T\mathbf{B}^T$. However note that we can form the product $\mathbf{B}^T\mathbf{A}^T$. One property of matrices is that $(\mathbf{AB})^T = \mathbf{B}^T\mathbf{A}^T$ (see the list of properties given in the last section of this appendix).

The Adjoint

The adjoint matrix of a square matrix **A** is formed by replacing each element a_{ij} by its cofactor A_{ij} and transposing the resulting matrix. Hence the adjoint of A is

$$\text{adj}(\mathbf{A}) = \begin{bmatrix} A_{11} & A_{12} & \cdots & A_{1n} \\ A_{21} & A_{22} & \cdots & A_{2n} \\ \vdots & \vdots & & \vdots \\ A_{n1} & A_{n2} & \cdots & A_{nn} \end{bmatrix}^T = \begin{bmatrix} A_{11} & A_{21} & \cdots & A_{n1} \\ A_{12} & A_{22} & \cdots & A_{n2} \\ \vdots & \vdots & & \vdots \\ A_{1n} & A_{2n} & \cdots & A_{nn} \end{bmatrix}.$$

For example, consider the matrix

$$\mathbf{B} = \begin{bmatrix} 1 & 3 & 2 \\ 6 & 1 & 2 \\ 3 & 4 & 1 \end{bmatrix}.$$

The adjoint of **B** is given by

$$\mathrm{adj}\,(\mathbf{B}) = \begin{bmatrix} -7 & 5 & 4 \\ 0 & -5 & 10 \\ 21 & 15 & -17 \end{bmatrix}.$$

The Inverse

The inverse of a square matrix **A** (denoted by $\mathbf{A}^{-1}$) is defined by

$$\mathbf{A}\mathbf{A}^{-1} = \mathbf{A}^{-1}\mathbf{A} = \mathbf{I},$$

where **I** is the identity matrix

$$\mathbf{I} = \begin{bmatrix} 1 & 0 & 0 & \cdots & 0 \\ 0 & 1 & 0 & \cdots & 0 \\ 0 & 0 & 1 & \cdots & 0 \\ \cdot & \cdot & \cdot & & \cdot \\ \cdot & \cdot & \cdot & & \cdot \\ \cdot & \cdot & \cdot & & \cdot \\ 0 & 0 & 0 & \cdots & 1 \end{bmatrix}.$$

A matrix is diagonal if all its entries in the off-diagonal position are zero. A diagonal matrix with each diagonal element equal to one is called the identity matrix **I**.

Note. The inverse of a matrix is analogous to the inverse of a number. For example, the inverse of 2 is $\frac{1}{2}$ because $2 \cdot \frac{1}{2} = \frac{1}{2} \cdot 2 = 1$.

The matrix **A** (and, therefore, $\mathbf{A}^{-1}$) must be square ($n \times n$), and the determinant $|\mathbf{A}|$ must not be equal to zero. The inverse of **A** is given by

$$\mathbf{A}^{-1} = \frac{\mathrm{adj}\,(\mathbf{A})}{|\mathbf{A}|}.$$

Examples

1. Find the inverse of **A** given by

$$\mathbf{A} = \begin{bmatrix} 3 & 4 \\ 1 & 2 \end{bmatrix}.$$

First check to see whether **A** is square and $|\mathbf{A}| \neq 0$. Here $|\mathbf{A}| = 6 - 4 = 2$. The adjoint of **A** is

$$\mathrm{adj}\,(\mathbf{A}) = \begin{bmatrix} 2 & -4 \\ -1 & 3 \end{bmatrix},$$

so the inverse is

$$\mathbf{A}^{-1} = \frac{\mathrm{adj}\,(\mathbf{A})}{|\mathbf{A}|} = \frac{\begin{bmatrix} 2 & -4 \\ -1 & 3 \end{bmatrix}}{2} = \begin{bmatrix} 1 & -2 \\ -\frac{1}{2} & \frac{3}{2} \end{bmatrix}.$$

By multiplying, check that $AA^{-1} = I$.

$$AA^{-1} = \begin{bmatrix} 3 & 4 \\ 1 & 2 \end{bmatrix} \begin{bmatrix} 1 & -2 \\ -\frac{1}{2} & \frac{3}{2} \end{bmatrix} = \begin{bmatrix} 1 & 0 \\ 0 & 1 \end{bmatrix}.$$

2. As another example, find the inverse of **B** given by

$$B = \begin{bmatrix} 1 & 3 & 2 \\ 6 & 1 & 2 \\ 3 & 4 & 1 \end{bmatrix}.$$

The determinant of **B** was found in the section on determinants to be 35. In the section on the adjoint we found that

$$\text{adj}(B) = \begin{bmatrix} -7 & 5 & 4 \\ 0 & -5 & 10 \\ 21 & 5 & -17 \end{bmatrix}.$$

Hence the inverse is given by

$$B^{-1} = \frac{\text{adj}(B)}{|B|} = \begin{bmatrix} -\frac{7}{35} & \frac{1}{7} & \frac{4}{35} \\ 0 & -\frac{1}{7} & \frac{2}{7} \\ \frac{3}{5} & \frac{1}{7} & -\frac{17}{35} \end{bmatrix}.$$

Checking, we find that $BB^{-1} = B^{-1}B = I$.

Properties of Matrices

1. The transpose of A^T is A: $(A^T)^T = A$.
2. The inverse of A^{-1} is A: $(A^{-1})^{-1} = A$.
3. The transpose and inverse operations are commutable; that is, $(A^{-1})^T = (A^T)^{-1}$.
4. $(AB)^T = B^T A^T$.
5. $(AB)^{-1} = B^{-1}A^{-1}$. (Here we must have $|A| \neq 0$, $|B| \neq 0$.)
6. If X and Y are vectors, A has an inverse, and $Y = AX$, then $X = A^{-1}Y$.
7. The determinant of the product of two square matrices is the product of the determinants: $|AB| = |A||B|$.

PART 3

Stochastic Processes and Random Signals

9
STOCHASTIC PROCESSES

9.1 DEFINITION OF A STOCHASTIC PROCESS

In discussing Fourier series and transforms, we found the concept of function useful. In discussing probability and random variables, again we found the concept of function useful. Now the underlying theme in our discussion of stochastic processes will once more be the concept of function.

Recall that in our attempts to apply the theory of probability and random variables to electrical waveforms we encountered a serious difficulty (Section 6.3): there was no way to account for the time dependence of the waveforms. Time had to be stopped, so to speak, at a particular (but arbitrary) value, say t_1. Then we could define a random variable whose range was the possible values of the waveform. This concept is now extended so that time dependence can be accounted for by the introduction of stochastic processes.

Definition 1. *Stochastic process.* A stochastic process $X(t, \zeta)$ is a function of two variables, t and ζ, with ζ an element of the sample space. In addition, for fixed t the function $X(t, \zeta)$ must satisfy the definition of random variable (Definition 1, Chapter 6).

Notes. a) Now instead of a random variable $X(\zeta)$, we have a function of two variables $X(t, \zeta)$. Although the parameter t can be arbitrary, we will identify it with time.

b) The word "stochastic" is derived from the Greek *stochastikos*, meaning random. There is nothing random about a stochastic process, just as there is nothing random about a random variable. The randomness arises from the experiment.

There are four possible interpretations of $X(t, \zeta)$, depending on the nature of t and ζ. (1) If t and ζ are fixed, then $X(t, \zeta)$ is a number. (2) If t and ζ are both variable, then $X(t, \zeta)$ is a stochastic process. (3) If t is fixed and ζ is variable, then $X(t, \zeta)$ is a random variable. Note that this is what we did in Chapter 6 when we discussed the application of random variables to voltage waveforms; we fixed time at the value t_1 and interpreted the measurement $v(t_1)$ as an element in the range of the random variable. Thus this corresponded to a stochastic process with t fixed.

The interpretation (4) with ζ fixed and t variable is of most interest to us, and deserves some elaboration. In fact, most of the remainder of this text will be concerned with this case.

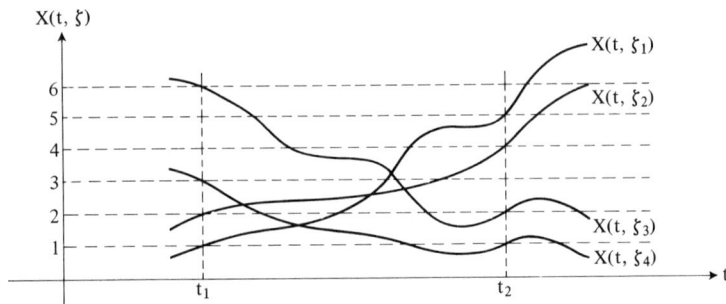

Fig. 9.1 The stochastic process $X(t, \zeta)$.

Consider Fig. 9.1. Four time functions are shown and identified with the four elements of the sample space, ζ_1, ζ_2, ζ_3, and ζ_4. We can think of performing the experiment so that the outcome is ζ_i. We then associate a time function, $X(t, \zeta_i)$ with the particular experimental outcome ζ_i. Thus we have a procedure for selecting a time function, and the particular function chosen depends on the experimental outcome ζ_i.

Notes. a) A set or collection of time functions such as that shown in Fig. 9.1 is called an *ensemble*. In this particular case the ensemble is finite, but in general we can deal with an infinite number of time functions.

b) We mention again that the interpretation of t as time is a special application. t can be identified with any appropriate physical attribute, or it can be constant (in which case we are dealing with a random variable). In fact, the term "ensemble" is often used in application to random variables, in which case we have an ensemble of numbers.

With reference to Fig. 9.1, if the probability law is known [that is, if $P(\zeta_1)$, $P(\zeta_2)$, $P(\zeta_3)$ and, $P(\zeta_4)$ are known], then the probability of selecting any one member of the ensemble is known. Also, if $P(\zeta_i)$ is known, then f_X and F_X are known. For the case of a continuous sample space, the probability of any one experimental outcome is zero, but we will be interested in computing average values (such as power) rather than individual probabilities. A knowledge of the cdf and pdf will allow us to compute these averages.

9.2 MOMENTS OF STOCHASTIC PROCESSES

The cdf and pdf of a stochastic process are, in general, time-dependent. Of course, these functions can be constant with time, but this is a special case. Thus instead of writing $F_X(x)$ for the cdf, as we do for a random variable, we write $F_X(x; t)$ to denote the time dependence. Similarly, we write $f_X(x; t)$ for the pdf.

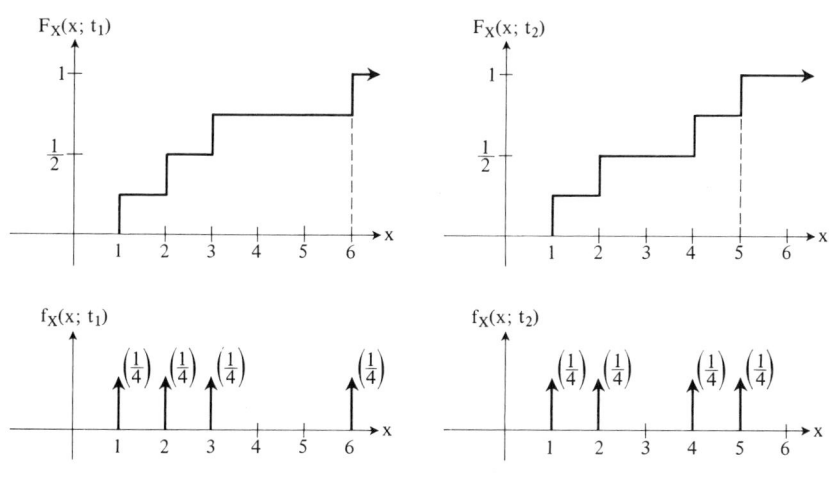

Fig. 9.2

Refer to Fig. 9.1 and suppose that $P(\zeta_i) = 0.25$ for $i = 1, 2, 3, 4$. The cdf's $F_X(x; t_1)$ and $F_X(x; t_2)$ are plotted in Fig. 9.2, along with the pdf's $f_X(x; t_1)$ and $f_X(x; t_2)$.

For a specific value of t, $X(t, \zeta)$ is a random variable. We can therefore speak of the cdf and pdf as in Chapter 6, and these functions are now time-dependent. The cdf and pdf are defined by

$$F_X(x; t) = P\{X(t) \leq x\}, \tag{9.1}$$

$$f_X(x; t) = \frac{d}{dx} F_X(x; t). \tag{9.2}$$

The definition of moments given in Section 6.4 applies here. Of course, these moments are now functions of t. The nth moment of $X(t, \zeta)$ is given by

$$m_n(t) = E(X^n) = \int_{-\infty}^{+\infty} x^n f_X(x; t)\, dx. \tag{9.3}$$

All this is well and good, but so far we have done nothing more with our new concept of stochastic process than we were able to do with the concept of random variable. If this were the complete story, then we would have added nothing to the concept of random variable. We have already applied the above concepts to electrical waveforms in Chapter 6, and the only difference is that we have now added the parameter t to the argument of $X(t, \zeta)$. Note that the important difference between stochastic processes and random variables is that now we have the entire waveform. Given an experimental outcome ζ, an entire waveform $X(t, \zeta)$ is specified, not just a number, as in the case of random variables. We should strive to make use of this additional information, and an important step in this direction is the use of joint moments, specifically the autocorrelation function.

In order to introduce joint moments for stochastic processes, we will follow the procedure given in Section 7.4 and first introduce the joint cdf and pdf. Given a stochastic process $X(t, \zeta)$ and two time instants t_1 and t_2, consider the random variables $X(t_1, \zeta)$ and $X(t_2, \zeta)$. The joint cdf and pdf are

$$F_{XX}(x_1, x_2; t_1, t_2) = P\{X(t_1) \leq x_1, X(t_2) \leq x_2\}, \qquad (9.4)$$

$$f_{XX}(x_1, x_2; t_1, t_2) = \frac{\partial^2}{\partial x_1, \partial x_2} F_{XX}(x_1, x_2; t_1, t_2). \qquad (9.5)$$

For example, refer to Fig. 9.1. The chart in Fig. 9.3(a) lists the possible values of $X(t_1)$ and $X(t_2)$ corresponding to the experimental outcomes ζ_i. The joint pdf and cdf for the times t_1 and t_2 are plotted in Fig. 9.3(b) and (c). These graphs would change if either t_1 or t_2 were changed.

Definition 2. *Autocorrelation function* $R_X(t_1, t_2)$. The autocorrelation function $R_X(t_1, t_2)$ of a stochastic process $X(t, \zeta)$ is the joint moment of the random variables $X(t_1, \zeta)$ and $X(t_2, \zeta)$:

$$R_X(t_1, t_2) = E[X(t_1, \zeta)X(t_2, \zeta)]$$

$$= \int_{-\infty}^{+\infty} \int_{-\infty}^{+\infty} x_1 x_2 f_{XX}(x_1, x_2; t_1, t_2) \, dx_1 \, dx_2. \qquad (9.6)$$

Notes. a) The domain of this function is two values of time, t_1 and t_2. (More specifically, the domain is the cross product $T \times T$, where T is the set of all values of time.) The range is a set of real numbers, usually $-\infty < R < +\infty$.

b) The subscript X on R_X denotes that this is the autocorrelation function for the stochastic process $X(t, \zeta)$.

c) If $t_1 = t_2$, then $R_X(t_1, t_1) = m_2(t_1)$, the second moment (mean-square value) of the process evaluated at $t = t_1$.

d) The value of $R_X(t_1, t_2)$ for the stochastic process shown in Fig. 9.1 is $\frac{1}{4}(5 + 8 + 12 + 3) = 7$.

9.2 MOMENTS OF STOCHASTIC PROCESSES

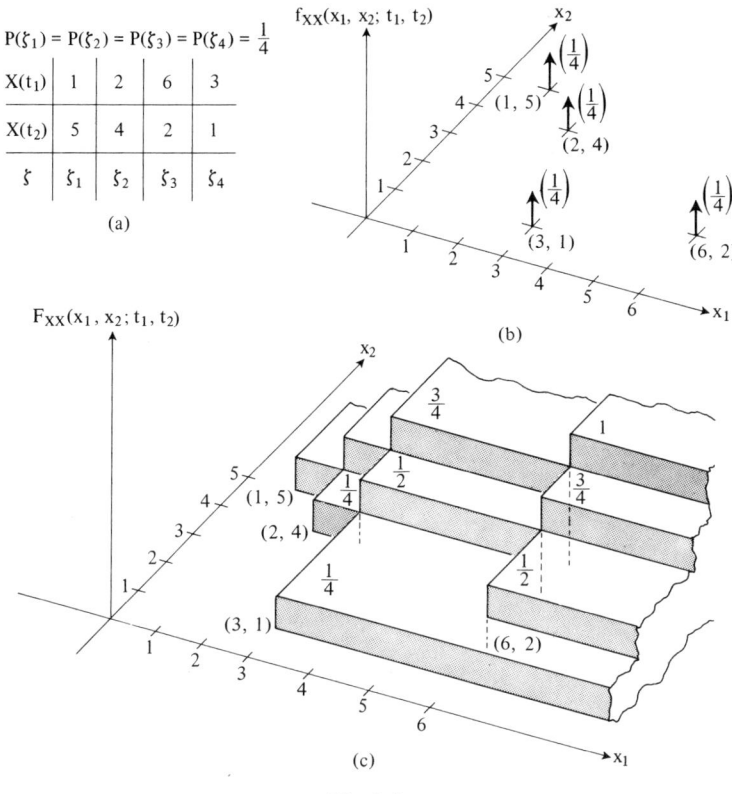

Fig. 9.3

Example 1. A sample function from a stochastic process is given by

$$x(t) = At, \quad -\infty < t < +\infty,$$

where A is a random variable with distribution f_A, as shown in Fig. 9.4. Find

a) the mean value $m_X(t)$,
b) the autocorrelation function $R_X(t_1, t_2)$.

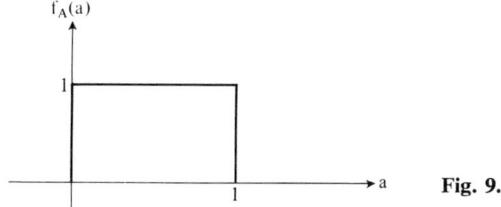

Fig. 9.4

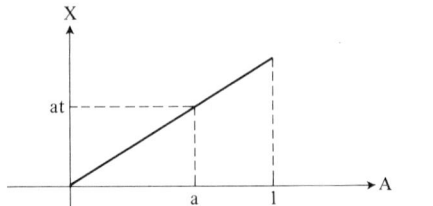

Fig. 9.5

Solution

a) The mean value is found by

$$m_X(t) = E[X(t)] = \int_{-\infty}^{+\infty} x f_X(x;t)\, dx. \tag{9.7}$$

However, f_X is unknown. To solve the problem here we will find f_X. Note that f_A and the relation between A and X are known. Hence we can use the procedures given in Section 7.2 to find the distribution of X to use in Eq. (9.7).

Another method of attack is to notice that the random variable $A(\zeta)$ and the stochastic process $X(t, \zeta)$ are defined on the same sample space. A particular experimental outcome ζ determines a particular value of A and also a particular function $X(t)$. Therefore equivalent events in R_A and R_X can be found and used to derive another expression for $m_X(t)$. This other expression will involve the integral of f_A and should therefore simplify matters in many cases. This procedure will be investigated in detail in Section 10.2.

Consider $t > 0$. The relationship between A and X is a straight line with slope t, as shown in Fig. 9.5. Thus the events $\{A \le a\}$ and $\{X \le at\}$ are equivalent, and their probabilities must be equal:

$$P\{X \le at\} = F_X(at) = F_A(a).$$

Therefore X is uniform $(0, t)$, and

$$m_X(t) = \int_0^t (1/t) x\, dx = t/2.$$

Similar reasoning for $t < 0$ leads to the same result. Therefore

$$m_X(t) = t/2, \qquad -\infty < t < +\infty. \tag{9.8}$$

b) The power of the second method becomes apparent in the evaluation of second- and higher-order moments. We must find either the second-order pdf or an expression equivalent to Eq. (9.6) that involves the first-order pdf f_A. We do not have at hand the tools necessary to find the second-order pdf $f_{XX}(x_1, x_2; t_1, t_2)$, though we could develop them without much effort. We do have at hand the tools for finding an expression equivalent to Eq. (9.6), and we will do this in Section 10.2. At that time we will find that

$$R_X(t_1, t_2) = \tfrac{1}{3} t_1 t_2, \qquad -\infty < t_1 < +\infty, \quad -\infty < t_2 < +\infty. \tag{9.9}$$

9.3 SPECIFYING STOCHASTIC PROCESSES

Consider a stochastic process $X(t, \zeta)$. The experiment is performed with outcome ζ, which in turn determines the particular sample function $X(t)$. Now let us suppose that k samples are taken, so that we have a random vector

$$\mathbf{X} = [X(t_1), X(t_2), \ldots, X(t_k)].$$

This random vector has a joint cdf $F_\mathbf{X}$ and a joint pdf $f_\mathbf{X}$. We now define the important concept of specifying a stochastic process.

Definition 3. *Specification of a stochastic process.* A stochastic process is specified if $F_\mathbf{X}$ can be determined for every finite set of time instants $t_1, t_2, \ldots, t_k$.

Note. This means that if *any* finite set of time instants (say 1000 of them) is specified, then we must be able to write down $F_\mathbf{X}$, by no means a trivial undertaking in the general case.

Methods of Specifying Stochastic Processes

1. State the rule for determining $F_\mathbf{X}$ directly. Of course, it is a special case when this is possible. For this to be practical the joint cdf must depend on the time instants in a known elementary way. Fortunately, the gaussian process can be specified in this manner.

2. *Transformation of variables.* A time function involving one or more parameters is given, where the parameters are random variables. Example 1 above is a case in point. The joint cdf $F_\mathbf{X}$ can be determined from knowledge of the cdf F_A given in the example. Another example that is quite common in the literature is that of a stochastic process defined by

$$X(t) = \gamma \cos(2\pi t + \theta), \tag{9.10}$$

where γ and θ are jointly distributed random variables with cdf $F_{\gamma,\theta}$. The cdf $F_\mathbf{X}$ can be determined from the functional relationship given in Eq. (9.10) and a knowledge of $F_{\gamma,\theta}$ by the procedures of Sections 7.1, 7.2, and 7.3.

3. A third method of specifying a stochastic process is by applying a stated operation to the sample functions of a known process. Here we are thinking specifically of systems. If the input is a sample function from a stochastic process and the mathematical model for the system is an operator, then the system output is a sample function from another stochastic process. In some cases we may be able to specify the output from a knowledge of the input and system transformation. A classic example is a linear operation performed on a gaussian process. In this case the output is gaussian, and we will see that it is a simple matter to specify the output when the input is specified.

9.4 THE GAUSSIAN PROCESS

We studied the guassian random variable in Sections 6.5 and 8.2. Now we will extend these concepts to stochastic process. Recall that an n-dimensional random variable is gaussian if each of the components is marginally gaussian. Similarly, a gaussian stochastic process is defined as follows:

Definition 4. *Gaussian stochastic process.* A stochastic process is called guassian if F_X is gaussian for every finite set of time instants $t_1, t_2, \ldots, t_k$.

Note. This means that the joint cdf (and pdf) of the random vector $\mathbf{X} = [X(t_1), X(t_2), \ldots, X(t_k)]$ is of the form of Eq. (8.23). For a different set of time instants the cdf (and pdf) may change, but they are always gaussian.

The Covariance and Correlation Functions

The gaussian process is one that can be specified by stating the rule directly. Recall that the pdf of the gaussian random variable is known if the first two moments (means, variances, and covariances) are known. Therefore, if a rule can be stated to determine these moments for any set of samples, the gaussian process is specified.

This rule can be stated in terms of the mean function and covariance or correlation function. The mean function describes the mean as a function of time. It is given by

$$m_X(t) = E[X(t)] = \int_{-\infty}^{+\infty} x(t) f_X(x; t) \, dx. \tag{9.11}$$

[This is Eq. (9.3) with $n = 1$.]

The covariance function $\mathfrak{L}_X(t_1, t_2)$ is related to the correlation function $R_X(t_1, t_2)$ of Eq. (9.6) by

$$\mathfrak{L}_X(t_1, t_2) = E\{[X(t_1, \zeta) - m_X(t_1)][X(t_2, \zeta) - m_X(t_2)]\}$$
$$= R_X(t_1, t_2) - m_X(t_1) m_X(t_2). \tag{9.12}$$

If the mean function and either the correction or the covariance function is given, then the mean vector $\mathbf{\bar{X}} = [\bar{X}(t_1), \bar{X}(t_2), \ldots, \bar{X}(t_k)]$ and the covariance matrix Λ_X can be determined in the guassian pdf. Thus the gaussian process is specified if $m_X(t)$ and $\mathfrak{L}_X(t_1, t_2)$ are known.

Example 2. Determine an expression for f_X for three samples taken from a gaussian process at times $t_1 = 0$, $t_2 = 1$, and $t_3 = 2$ sec. The mean is zero and the covariance function is

$$\mathfrak{L}_X(t, s) = e^{-|s-t|}. \tag{9.13}$$

Solution. The pdf $f_X(\mathbf{x})$ is of the form

$$f_X(\mathbf{x}) = \frac{1}{(2\pi)^{3/2} |\Lambda_X|^{1/2}} \exp\left[-\tfrac{1}{2} \mathbf{x} \Lambda_X^{-1} \mathbf{x}^T\right].$$

We need to determine only the covariance matrix Λ_X:

$$\Lambda_X = \begin{bmatrix} \mathfrak{L}_X(t_1, t_1) & \mathfrak{L}_X(t_1, t_2) & \mathfrak{L}_X(t_1, t_3) \\ \mathfrak{L}_X(t_2, t_1) & \mathfrak{L}_X(t_2, t_2) & \mathfrak{L}_X(t_2, t_3) \\ \mathfrak{L}_X(t_3, t_1) & \mathfrak{L}_X(t_3, t_2) & \mathfrak{L}_X(t_3, t_3) \end{bmatrix}$$

$$= \begin{bmatrix} E(X_1^2) & E(X_1 X_2) & E(X_1 X_3) \\ E(X_2 X_1) & E(X_2^2) & E(X_2 X_3) \\ E(X_3 X_1) & E(X_3 X_2) & E(X_3^2) \end{bmatrix},$$

where we have simplified the notation by using $X_1 = X(t_1)$, $X_2 = X(t_2)$, and $X_3 = X(t_3)$.

Substituting values for t and s in Eq. (9.13), we find that the covariance matrix is given by

$$\Lambda_X = \begin{bmatrix} 1 & e^{-1} & e^{-2} \\ e^{-1} & 1 & e^{-1} \\ e^{-2} & e^{-1} & 1 \end{bmatrix}.$$

Substitution of this covariance matrix into the above expression for $f_\mathbf{X}(\mathbf{x})$ then completes the solution.

9.5 STATIONARY AND ERGODIC PROCESSES

In practice we are often faced with a situation in which one member function from a stochastic process is available to us. In order to use a statistical model in our analysis, it is necessary to derive all statistical information from this one member function. Furthermore, the statistical information gained at one time should be valid at some later time (the next day, or next month).

In order for both of these conditions to hold, the process must be stationary and ergodic. Stationary implies that the statistics are invariant with time. Ergodicity implies that all statistical information can be gained from one member function.

In the following discussion we shall write $X(t)$ in place of $X(t, \zeta)$. The dependence on ζ is understood.

A stochastic process is stationary if the statistics of the random variables $X(t_1)$ and $X(t_2)$ are equal. That is, the statistics determined for $X(t)$ are equal to those for $X(t + \epsilon)$ for every ϵ.

Definition 5. *Stationarity.* A stochastic process is stationary of order K if

$$f_\mathbf{X}(x_1, \ldots, x_K; t_1, \ldots, t_K) = f_\mathbf{X}(x_1, \ldots, x_K; t_1 + \epsilon, \ldots, t_K + \epsilon)$$

for all ϵ. (9.14)

Since there are several types of stationarity, some special terminology has arisen. A process is called *strictly stationary* if it is stationary for any order, $K = 1, 2, \ldots$ A process is called *wide-sense* (or weakly) *stationary* if its mean value is a constant and its autocorrelation depends only on $\tau = t_2 - t_1$. Note that this is not the same as stationarity of order two. Any second-order stationary process is wide-sense stationary, but the converse is not always true.

Consider a second-order stationary process. It must be true that

$$f_X(x; t) = f_X(x; t + \epsilon) \quad \text{for all } \epsilon.$$

Thus the first-order density function must be independent of t. Since the mean value is a functional that depends on (only) the first-order density function, the mean must be independent of t:

$$E[X(t)] = m_X(t) = m_X = \text{constant}.$$

For a second-order stationary process it must also be true that

$$f_X(x_1, x_2; t_1, t_2) = f_X(x_1, x_2; t_1 + \epsilon, t_2 + \epsilon) \quad \text{for all } \epsilon,$$

which is a function only of $\tau = t_2 - t_1$. Now the autocorrelation function is a functional* that depends on the second-order density function, so that

$$R_X(t_1, t_2) = R_X(t_2 - t_1) = R_X(\tau).$$

Thus for a second- (or higher-) order stationary process, the autocorrelation function is dependent only on $\tau = t_2 - t_1$.

Properties of $R_X(\tau)$

For a stationary process the autocorrelation function has the following properties:

1. $R_X(0) \geq |R_X(\tau)|$.
2. $R_X(0) = E(X^2)$.
3. $R_X(\tau) = R_X(-\tau)$.

Properties 2 and 3 are obvious from the definition of $R_X(\tau)$. Property 1 is shown as follows: Consider

$$E\{[X(t + \tau) - X(t)]^2\} = 2[R_X(0) - R_X(\tau)]. \tag{9.15}$$

Since the left-hand side of Eq. (9.15) is greater than or equal to zero, property 1 follows.

---

*R_X is a functional so far as the experimental outcome is concerned. That is, for particular values t_1 and t_2 the pdf is mapped into the *number* $R_X(t_1, t_2)$.

Note. If $R_X(\tau)$ has a Fourier transform $S_X(f)$,

$$R_X(\tau) \leftrightarrow S_X(f),$$

with the property that $|S_X(f)| = S_X(f)$, then property 1 can be shown as follows:

$$|R_X(\tau)| = \left| \int_{-\infty}^{+\infty} S_X(f) e^{j2\pi f \tau} \, df \right|$$

$$\leq \int_{-\infty}^{+\infty} |S_X(f)| \, |e^{j2\pi f \tau}| \, df = \int_{-\infty}^{+\infty} |S_X(f)| \, df = R_X(0).$$

We will investigate the conditions under which $S_X(f)$ exists in Section 10.1.

Ergodicity

The notion of stationarity developed from the concept of time invariance. That is, the statistics of a stochastic process should be the same tomorrow as they are today. An ergodic process has this property, in addition to some other useful properties.

The idea of ergodicity arises if only one sample function from a stochastic process is available, instead of the entire ensemble. Usually a single sample function provides no information about the statistics of the process. However, if the process is ergodic, then *all* statistical information can be derived from just *one* sample function.

Definition 6. *Ergodicity.* A stochastic process is ergodic if time averages equal ensemble averages.

There are degrees of ergodicity, just as there are degrees of stationarity. We will discuss three degrees of ergodicity: ergodicity in the mean, in correlation, and in distribution.

A process is ergodic in the mean if

$$\lim_{T \to \infty} \frac{1}{2T} \int_{-T}^{+T} x(t) \, dt = E[X(t, \zeta)]. \tag{9.16}$$

Notes. a) We can compute the left-hand side of Eq. (9.16) by first selecting a particular member function $x(t)$ and then averaging in time. To compute the right-hand side, we must know the first-order pdf $f_X(x; t)$.

b) The left-hand side of Eq. (9.16) is a number. Hence the mean

$$m(t) = E[X(t, \zeta)]$$

must be a constant. Therefore, ergodicity of the mean implies stationarity of the mean.

c) Stationarity of the mean does not necessarily imply ergodicity of the mean, as our first example on page 194 indicates.

Examples

3. Consider a basket full of batteries. There are some flashlight batteries, some car batteries, and several other kinds of batteries. Suppose a battery is selected at random and its voltage is measured. This battery voltage $v(t)$ is a member function from a class of constant battery voltages, that is, a member function from a stochastic process.

This stochastic process is stationary but not ergodic in the mean. The time average is equal to the particular battery voltage selected (say 1.5 V). The statistical average $E[X(t, \zeta)]$ is some other number depending on what is in the basket. Thus Eq. (9.16) does not hold.

4. Let $x(t) = \cos(\omega t + \theta)$ be a member function from a stochastic process specified by a transformation of variables. If

$$f_\Theta(\theta) = \frac{1}{2\pi}, \quad 0 < \theta < 2\pi,$$

then the stochastic process is ergodic in the mean. For any other distribution of Θ the process is not stationary in the mean and hence not ergodic.

5. The periodic square wave shown in Fig. 9.6 has random starting time t_0 uniformly distributed over one period:

$$f_{T_0}(t_0) = \frac{1}{T}, \quad 0 < t_0 < T.$$

Since this waveform can be expressed as the sum of sinusoids, as in the above example, the process is ergodic. Again, for any other distribution of T_0 the process is not stationary.

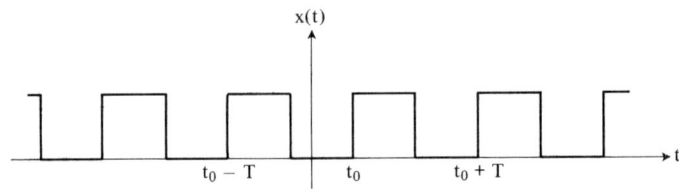

Fig. 9.6

The second degree of ergodicity is concerned with the autocorrelation function. A process is ergodic in correlation if

$$\lim_{T \to \infty} \frac{1}{2T} \int_{-T}^{+T} x(t)x(t + \tau)\, dt = E[X(t)X(t + \tau)]. \tag{9.17}$$

Notes. a) Again as in Eq. (9.16), we can compute the left-hand side by using a particular function $x(t)$. To compute the right-hand side, we must know the second-order pdf.

b) The left-hand side of Eq. (9.17) is a function of τ only. Hence if a process is ergodic in both the mean and correlation, it must be wide-sense stationary.

c) Wide-sense stationarity does not imply ergodicity of any order.

Consider again Examples 3, 4, and 5. The basket full of batteries is wide-sense stationary (and more) but not ergodic in any sense. The processes of Examples 4 and 5 are ergodic in correlation (and more), and therefore are wide-sense stationary.

The third degree of ergodicity is ergodicity in distribution. In this case the first-order distribution function $F_X(x; t)$ can be calculated from a single sample function. Figure 9.7 illustrates a member function from a stochastic process. The probability $F_X(x; t) = P\{X(t) \leq x\}$ is calculated for the particular value of x shown by the ratio of the time the function is less than x over the total time $2T$:

$$P\{X \leq x\} = 1 - \frac{l_1 + l_2}{2T} \qquad (9.18)$$

in the time interval $-T < t < +T$. The cdf is then the limit of such an expression as $T \to \infty$.

Note. Again we have used a single member function to compute a time average, the right-hand side of Eq. (9.18).

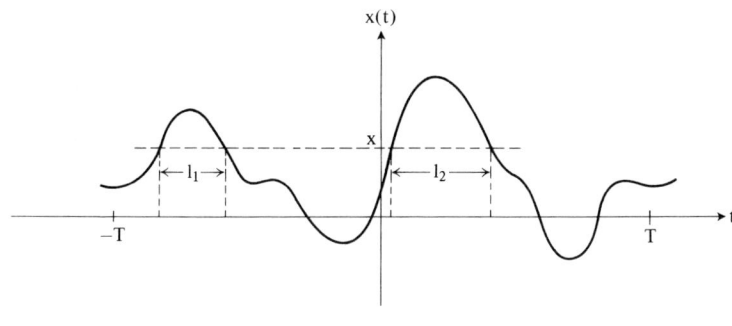

Fig. 9.7

PROBLEMS

1. Consider a stochastic process that has only the four distinct outcomes illustrated in Fig. 9.8. The probabilities are

$$P(\zeta_1) = \tfrac{1}{8}, \qquad P(\zeta_2) = \tfrac{1}{4}, \qquad P(\zeta_3) = \tfrac{3}{8}, \qquad P(\zeta_4) = \tfrac{1}{4}.$$

Find

a) the pdf of $X(t)$ for $t = t_1$,
b) the pdf of $X(t)$ for $t = t_2$,

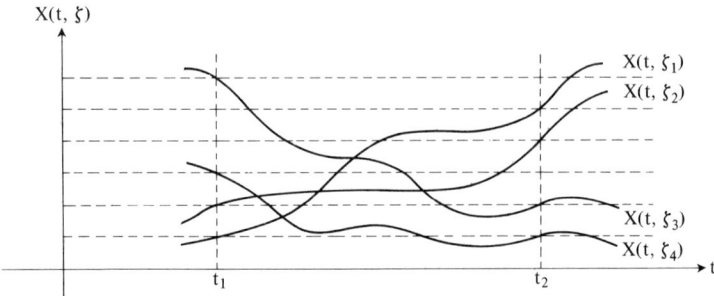

Fig. 9.8

c) $E[X(t_1)]$,
d) $E[X(t_2)]$,
e) $E[X(t_1)X(t_2)]$,
f) the joint density function $f_{X_1 X_2}(x_1, x_2; t_1, t_2)$.

2. A stochastic process is generated as follows: Starting at $t = 0$, we toss a coin every second. If heads comes up, the random variable X is set equal to one. If tails come up, $X = 0$. Hence a stochastic process is generated, and a typical member function is shown in Fig. 9.9. Calculate

a) the mean, $m_X(t)$,
b) the correlation, $R_X(0.5, 0.6)$,
c) the correlation, $R_X(0.5, 1.5)$.

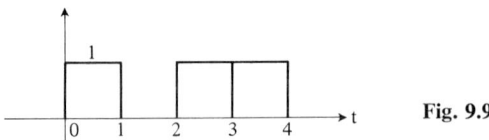

Fig. 9.9

3. A stochastic process consists of three functions of time, each occurring with equal probability as shown in Fig. 9.10. Calculate

$$E[X(2)], \quad E[X(6)], \quad F_X(x, 2),$$
$$F_X(x, 6), \quad F_{XX}(x_1, x_2; 2, 6), \quad R_{XX}(2, 6).$$

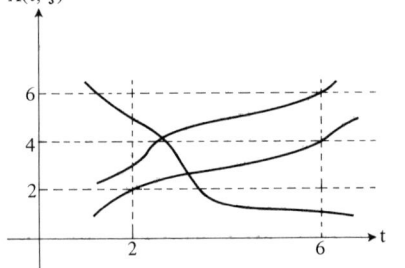

Fig. 9.10

4. A random process consists of three sample functions,

$$X(t, \zeta_1) = 1, \qquad X(t, \zeta_2) = \sin t, \qquad X(t, \zeta_3) = \cos t,$$

each occurring with equal probability. Is the process stationary? Why?

5. For the above process calculate
a) the mean $m_X(t)$,
b) the autocorrelation $R_X(t_1, t_2)$.

6. Two statistically independent, stationary random processes $X(t)$ and $Y(t)$ with zero mean have the following autocorrelation functions:

$$R_X(\tau) = e^{-|\tau|}, \qquad R_Y(\tau) = \cos 2\pi\tau.$$

Find
a) the autocorrelation function of $Z(t) = X(t) + Y(t)$,
b) the autocorrelation function of $W(t) = X(t) - Y(t)$,
c) the cross-correlation function $R_{ZW}(\tau)$.

7. A stochastic process is specified by a transformation of variables as

$$X(t) = A \cos(\omega t + \Theta),$$

where A and ω are constants but Θ is a random variable uniformly distributed between 0 and π. Determine whether the process is stationary.

8. A gaussian process has mean $m_X(t) = 2$ and covariance function $\mathcal{L}_X(t_1, t_2) = 8 \cos \pi(t_2 - t_1)$. Samples are taken at $t_1 = 0$ and $t_2 = \frac{1}{2}$. Write the second-order density function.

9. A stationary gaussian random process with zero mean has autocorrelation

$$R_X(\tau) = \frac{\sin \pi\tau}{\pi\tau}.$$

Write the third-order density function for samples taken at

$$t_1 = 0, \qquad t_2 = \tfrac{1}{2}, \qquad t_3 = 1.$$

10. The stochastic process $X(t)$ is stationary. Show that if for large τ the random variables $X(t + \tau)$ and $X(t)$ are uncorrelated, then

$$\lim_{\tau \to \infty} R_X(\tau) = \{E[X(t)]\}^2.$$

11. The autocorrelation function of a stochastic process is shown in Fig. 9.11.
a) Can you determine the mean-square value of the process? If so, how?
b) Can you determine the average value (dc-component)? If so, how?

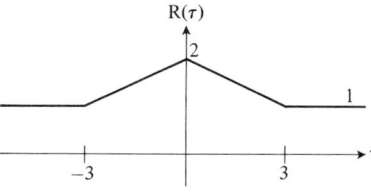

Fig. 9.11

c) Can you determine the variance (power in the ac-component)? How?
d) Can you determine whether the process is stationary, or even wide-sense stationary? If, so how?
e) Is the process periodic? Why (or why not)?

FURTHER READING

1. A. PAPOULIS, *Probability, Random Variables, and Stochastic Processes*, McGraw-Hill, New York, 1965.
2. J. M. WOZENCRAFT and I. M. JACOBS, *Principles of Communication Engineering*, Wiley, New York, 1965.

Both of these texts discuss all the material of this chapter. Papoulis is especially good on the material covered in Sections 9.1, 9.2, and 9.5. Wozencraft and Jacobs are better on the material discussed in Sections 9.3 and 9.4.

10
THE POWER SPECTRUM AND ITS RELATION TO THE AUTOCORRELATION FUNCTION

10.1 POWER SPECTRAL DENSITY AND AUTOCORRELATION

In Section 9.2 we defined the autocorrelation function as the expected value of the product of the random variables $X(t_1)$ and $X(t_2)$. We discussed the power spectral density in Section 2.7. Under certain conditions these two seemingly unrelated functions form a Fourier transform pair. In this section we will investigate those conditions.

Let $x(t)$ be a sample function from a stochastic process. Define the truncated function $x_T(t)$ by

$$x_T(t) = \begin{cases} x(t), & |t| \leq T, \\ 0, & |t| > T. \end{cases} \tag{10.1}$$

Note that

$$x(t) = \lim_{T \to \infty} x_T(t).$$

See Fig. 10.1.

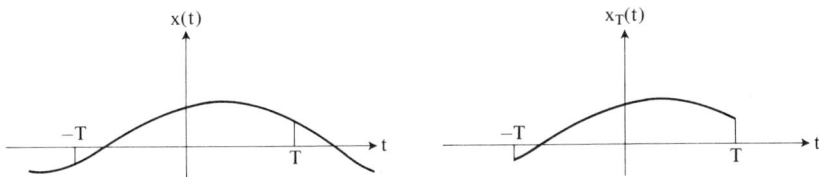

Fig. 10.1 The function $x(t)$ and the truncated function $x_T(t)$.

Note. This truncated function is defined so that we can take the Fourier transform of $x_T(t)$. Recall that the transform of a power signal is not defined. Thus if $x(t)$ is a power signal, we have

$$\int_{-\infty}^{+\infty} |x(t)|\, dt \not< \infty \quad \text{but} \quad \int_{-\infty}^{+\infty} |x_T(t)|\, dt < \infty.$$

Define the truncated Fourier transform pair as

$$X_T(f) = \int_{-\infty}^{+\infty} x_T(t) e^{-j\omega t} \, dt = \int_{-T}^{+T} x_T(t) e^{-j\omega t} \, dt, \qquad (10.2)$$

$$x_T(t) = \int_{-\infty}^{+\infty} X_T(f) e^{j\omega t} \, df. \qquad (10.3)$$

Now consider for a moment the physical significance of what we are doing. If $x(t)$ is a power signal, there must be a power spectral density function associated with it, and the total area under this density must be the average power. All this is true despite the fact that the Fourier transform of $x(t)$ does not exist. The definition of average power is given by Eq. (1.1). In the following sequence we begin with Eq. (1.1) and work backward to find the function of frequency whose integral yields the average power. This we define as the power spectral density function. The average power is given by

$$P = \lim_{T \to \infty} \frac{1}{2T} \int_{-T}^{+T} |x(t)|^2 \, dt$$

$$= \lim_{T \to \infty} \frac{1}{2T} \int_{-T}^{+T} |x_T(t)|^2 \, dt, \qquad (1.1)$$

since $x_T(t) = x(t)$ for $|t| < T$. Substituting Eq. (10.3), we get

$$P = \lim_{T \to \infty} \frac{1}{2T} \int_{-T}^{+T} x_T^*(t) \left[\int_{-\infty}^{+\infty} X_T(f) e^{j\omega t} \, df \right] dt$$

$$= \lim_{T \to \infty} \frac{1}{2T} \int_{-\infty}^{+\infty} X_T(f) \left[\int_{-T}^{+T} x_T^*(t) e^{j\omega t} \, dt \right] df$$

$$= \lim_{T \to \infty} \frac{1}{2T} \int_{-\infty}^{+\infty} |X_T(f)|^2 \, df$$

$$= \int_{-\infty}^{+\infty} \lim_{T \to \infty} \frac{1}{2T} |X_T(f)|^2 \, df.$$

Recall that in Section 2.7 the power spectral density function was described as that function of frequency which (1) gave the total power when integrated and (2) described the relative amount of power in each frequency band. The function of frequency inside the above integral does this, and therefore is called the power spectral density function for the signal $x(t)$. It is given by

$$G_X(f, \zeta) = \lim_{T \to \infty} \frac{1}{2T} |X_T(f, \zeta)|^2. \qquad (10.4)$$

Now Eq. (10.4) is for one member function (a particular experimental outcome ζ) and hence depends on ζ. The statistical average over all experimental outcomes is defined to be the power spectral density function of the process.

10.1 POWER SPECTRAL DENSITY AND AUTOCORRELATION

Definition 1. *Power spectral density.* Provided the limit exists, the power spectral density function for a stochastic process is the average

$$G_X(f) = E\left[\lim_{T \to \infty} \frac{1}{2T} |X_T(f, \zeta)|^2\right] = E[G_X(f, \zeta)]. \qquad (10.5)$$

Notes. a) The function $G_X(f, \zeta)$ of Eq. (10.4) is the power spectral density for one particular time function $x(t)$. The function $G_X(f)$ of Eq. (10.5) is (by definition) the power spectral density for the process.

b) The subscript X on $G_X(f)$ indicates that this is the density for the stochastic process $X(t, \zeta)$.

Relationship Between $G(f)$ and $R(\tau)$

The time autocorrelation function is given by

$$\mathcal{R}_X(\tau, \zeta) = \lim_{T \to \infty} \frac{1}{2T} \int_{-T}^{+T} x_T(t) x_T(t + \tau) \, dt.$$

Notes. a) $\mathcal{R}$ denotes the time autocorrelation function and R denotes the ensemble autocorrelation function.

b) If the process is ergodic, then $\mathcal{R}_X(\tau, \zeta) = R_X(\tau)$.

c) $\mathcal{R}_X(\tau, \zeta)$ is a function of the experimental outcome ζ, since $x(t)$ is determined by the experimental outcome.

Now we take the Fourier transform of $\mathcal{R}_X(\tau, \zeta)$ and get

$$\int_{-\infty}^{+\infty} \mathcal{R}_X(\tau, \zeta) e^{-j\omega\tau} \, d\tau = \int_{-\infty}^{+\infty} e^{-j\omega\tau} \left[\lim_{T \to \infty} \frac{1}{2T} \int_{-T}^{+T} x_T(t) x_T(t + \tau) \, dt\right] d\tau$$

$$= \lim_{T \to \infty} \frac{1}{2T} \int_{-T}^{+T} x_T(t) \left[\int_{-\infty}^{+\infty} x_T(t + \tau) e^{-j\omega\tau} \, d\tau\right] dt.$$

Multiplying by $e^{-j\omega t} e^{j\omega t} = 1$, we obtain

$$\lim_{T \to \infty} \frac{1}{2T} \int_{-T}^{+T} x_T(t) e^{j\omega t} \left[\int_{-\infty}^{+\infty} x_T(t + \tau) e^{-j\omega(t+\tau)} \, d\tau\right] dt.$$

We recognize that

$$\int_{-\infty}^{+\infty} x_T(t + \tau) e^{-j\omega(t+\tau)} \, d\tau = \int_{-\infty}^{+\infty} x_T(\tau) e^{-j\omega\tau} \, d\tau = X_T(f),$$

which can be shown by a simple change of variable. Also $x_T^*(t) = x_T(t)$ if the stochastic process is real (it always is here). Therefore

$$\int_{-\infty}^{+\infty} \mathcal{R}_X(\tau, \zeta) e^{-j\omega\tau} \, d\tau = \lim_{T \to \infty} \frac{1}{2T} \int_{-T}^{+T} x_T^*(t) e^{j\omega t} X_T(f, \zeta) \, dt$$

$$= \lim_{T \to \infty} \frac{1}{2T} |X_T(f, \zeta)|^2 = G_X(f, \zeta).$$

For a single function $x(t)$ from a stochastic process, the time autocorrelation function and the power spectral density function form a Fourier transform pair. Now we apply Eq. (10.5) to find the power spectral density for the process:

$$G_X(f) = E\left[\int_{-\infty}^{+\infty} \mathcal{R}_X(\tau, \zeta)e^{-j\omega\tau}\, d\tau\right]$$

$$= E\left\{\int_{-\infty}^{+\infty} e^{-j\omega\tau}\left[\lim_{T\to\infty}\frac{1}{2T}\int_{-T}^{+T} x(t)x(t+\tau)\, dt\right] d\tau\right\}$$

$$= \int_{-\infty}^{+\infty} e^{-j\omega\tau}\left\{\lim_{T\to\infty}\frac{1}{2T}\int_{-T}^{+T} E[x(t)x(t+\tau)]\, dt\right\} d\tau.$$

But $E[x(t)x(t+\tau)] = R_X(\tau)$ is a constant with respect to the variable of integration t if $X(t, \zeta)$ is wide-sense stationary. Then the term inside the braces reduces to $R_X(\tau)$, and

$$G_X(f) = \int_{-\infty}^{+\infty} R_X(\tau)e^{-j\omega\tau}\, d\tau.$$

Note. Although the process must be ergodic for $\mathcal{R}_X(\tau, \zeta) = R_X(\tau)$, wide-sense stationarity is the only requirement for $G_X(f) \leftrightarrow R_X(\tau)$.

In summary, for a *wide-sense stationary* process,

$$R_X(\tau) = \int_{-\infty}^{+\infty} G_X(f)e^{j2\pi f\tau}\, df, \qquad (10.6)$$

$$G_X(f) = \int_{-\infty}^{+\infty} R_X(\tau)e^{-j2\pi f\tau}\, d\tau, \qquad (10.7)$$

where $G_X(f)$ is the power spectral density of Eq. (10.5) and $R_X(\tau)$ is the autocorrelation function $R_X(\tau) = E[x(t)x(t+\tau)]$.

Note. We made a slight change in the notation above. We used lower-case $x(t)$ for both the single sample function and the stochastic process. We did this to avoid confusion with the Fourier transform $X(f)$.

10.2 CALCULATING MOMENTS OF STOCHASTIC PROCESSES SPECIFIED BY THE TRANSFORMATION OF VARIABLES

Recall that there are three methods of specifying stochastic processes: (1) by stating the rule for finding F_X directly (this is possible with gaussian processes), (2) by transformation of variables (where the stochastic process depends on parameters that are random variables with known cdf's), and (3) by operating on a specified process (systems). We will now study procedures for computing moments of stochastic processes (the mean and autocorrelation functions) for these three specification methods.

10.2 CALCULATING MOMENTS OF STOCHASTIC PROCESSES

The problem of evaluating moments for a stochastic process specified by the first method is trivial. Here a rule that allows one to write F_X for all finite samples is given. Hence, finding any order moment is straightforward. Our primary interest is with stochastic processes specified by the second and third methods. In this section we consider the problem in connection with transformation of variables. In Section 10.3 we will study system input-output relations.

Equivalent events are an important and fruitful concept in connection with finding moments, just as they were in the transformation of variables (Chapter 7). Some examples of stochastic processes specified by the second method are given below, and moments are calculated. The equivalent-event concept is used extensively.

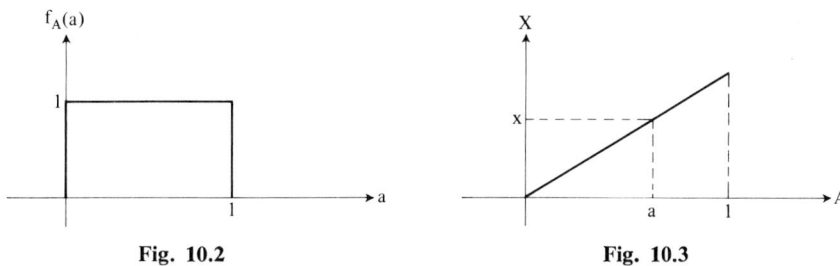

Fig. 10.2 Fig. 10.3

Examples

1. Consider again Example 1 of Section 9.2. A sample function from a stochastic process is given by

$$X(t) = At, \quad -\infty < t < +\infty,$$

where A is a random variable with the pdf f_A shown in Fig. 10.2 Find
a) the mean value $m_X(t)$,
b) the autocorrelation function $R_X(t_1, t_2)$.

Solution. a) As discussed in Section 9.2, we want to find the equivalent events in R_A and R_X, the range of the random variables A and X. The relationship between X and A is the straight line with slope t shown in Fig. 10.3. Consider first $t > 0$. The events $\{X \leq at\}$ and $\{A \leq a\}$ are equivalent for $0 < a < 1$. Therefore

$$F_X(at) = F_A(a). \tag{10.8}$$

From the relationship between the cdf and pdf we can write

$$dF_X(x) = f_X(x)\, dx \tag{10.9}$$

and

$$dF_A(a) = f_A(a)\, da. \tag{10.10}$$

Now from Eq. (10.9) the mean value is given by

$$m_X(t) = \int_{-\infty}^{+\infty} x f_X(x;t)\, dx = \int_{-\infty}^{+\infty} x\, dF_X(x;t),$$

where this last integral is a Lebesgue integral. (It is not necessary to know anything about the Lebesgue integral to understand our line of reasoning.) From the equality $x = at$ we can write

$$\int_{-\infty}^{+\infty} x\, dF_X(x) = \int_{-\infty}^{+\infty} (at)\, dF_X(at),$$

and because of Eq. (10.8) we can write

$$\int_{-\infty}^{+\infty} (at)\, dF_X(at) = \int_{-\infty}^{+\infty} (at)\, dF_A(a).$$

Finally, using Eq. (10.10), we get

$$\int_{-\infty}^{+\infty} (at)\, dF_A(a) = \int_{-\infty}^{+\infty} (at) f_A(a)\, da$$

or

$$m_X(t) = \int_{-\infty}^{+\infty} (at) f_A(a)\, da. \tag{10.11}$$

Note. From another viewpoint, X is a function of the random variable A. Equation (6.14) for the expected value of a function of a random variable is restated here in terms of X and A:

$$E[X(A)] = \int_{-\infty}^{+\infty} X(a) f_A(a)\, da$$

$$= \int_{-\infty}^{+\infty} (at) f_A(a)\, da,$$

which is Eq. (10.11). Therefore we can consider the stochastic process X to be a function of the random variable A and can use Eq. (6.14) to find the average value of this function of A. This is the viewpoint we will use to solve for the autocorrelation function.

b) Now to solve for $R_X(t_1, t_2)$, let us first reexamine the concept of using Eq. (6.14). The random variable A is defined on a sample space S with experimental outcomes ζ. For each ζ there corresponds a value of A, say a (Fig. 10.4). Each value a in turn determines a sample function $x(t) = at$, and thus for a particular time t the value of $x(t)$ is determined by the experimental outcome ζ.

We can work in the range space R_X and compute the average of X by

$$\int_{-\infty}^{+\infty} x f_X(x)\, dx,$$

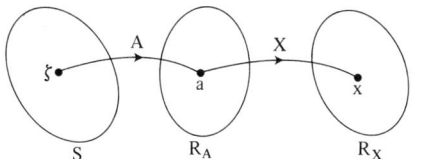

Fig. 10.4

or we can work in the range space R_A and compute the average of the function $X(A)$ by

$$\int_{-\infty}^{+\infty} X(a) f_A(a)\, da.$$

These two expressions are equivalent.

Consider the autocorrelation function

$$R_X(t_1, t_2) = E[X(t_1)X(t_2)].$$

This product is also a function of the random variable A; that is,

$$X(t_1)X(t_2) = [X(t_1)X(t_2)](A),$$

or for fixed t_1 and t_2 the value of the product is a function of the random variable A (Fig. 10.5). Thus we can calculate the average value of this product by using Eq. (6.14), and we get

$$R_X(t_1, t_2) = \int_{-\infty}^{+\infty} X(t_1)X(t_2) f_A(a)\, da$$

$$= \int_0^1 a^2 t_1 t_2\, da = \tfrac{1}{3} t_1 t_2.$$

Note. The functions $X(t)$ and $X(t_1)X(t_2)$ in parts (a) and (b) of this example are functions of *one* random variable. We will later encounter functions of two random variables, and this procedure will be modified accordingly.

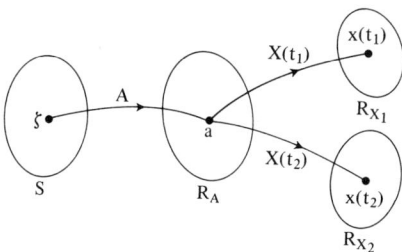

Fig. 10.5

2. Another example of a stochastic process specified by transformation of variables is the function

$$X(t) = p \cos(\omega t + \Theta), \tag{10.12}$$

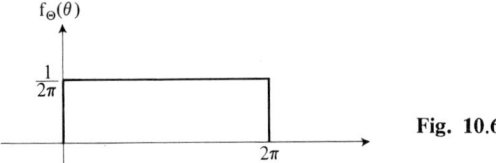

Fig. 10.6

where Θ is a random variable uniformly distributed $(0, 2\pi)$, as shown in Fig. 10.6. Compute the mean value and the autocorrelation function.

Solution. The stochastic process $X(t)$ is a function of a random variable, so again we use Eq. (5.14) to compute $m_X(t)$ and $R_X(t_1, t_2)$. Thus we get

$$m_X(t) = \int_{-\infty}^{+\infty} p \cos(\omega t + \theta) f_\Theta(\theta)\, d\theta$$

$$= \int_0^{2\pi} p \cos(\omega t + \theta) \frac{1}{2\pi}\, d\theta = 0$$

and

$$R_X(t_1, t_2) = \int_{-\infty}^{+\infty} p^2 \cos(\omega t_1 + \theta) \cos(\omega t_2 + \theta) f_\Theta(\theta)\, d\theta$$

$$= \frac{1}{2\pi} \int_0^{2\pi} p^2 \cos(\omega t_1 + \theta) \cos(\omega t_2 + \theta)\, d\theta$$

$$= \frac{p^2}{2} \cos(\omega t_1 - \omega t_2)$$

$$= \frac{p^2}{2} \cos \omega \tau \quad \text{with} \quad \tau = t_1 - t_2.$$

Note. The stochastic process of Eq. (10.12) is wide-sense stationary (at least). The mean is constant with time and the autocorrelation function depends on only the time difference $t_1 - t_2 = \tau$.

3. Now let us consider the case in which the stochastic process is a function of two random variables. An example of such a function is

$$X(t) = P \cos(\omega t + \Theta), \qquad (10.13)$$

where P and Θ are statistically independent random variables with distributions

$$f_P(p) = 1, \qquad 0 < p < 1,$$

$$f_\Theta(\theta) = \frac{1}{2\pi}, \qquad 0 < \theta < 2\pi.$$

A particular experimental outcome ζ determines a value p and a value θ (Fig. 10.7). These two values (p, θ) are in the domain of the stochastic

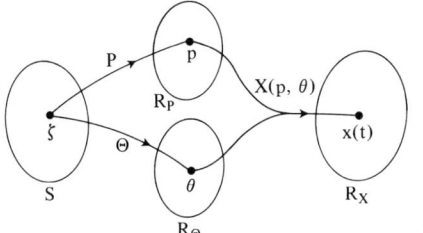

Fig. 10.7

process X, so the ordered pair (p, θ) determines a particular $x(t)$. Therefore the mean is given by

$$m_X(t) = \int_{-\infty}^{+\infty} \int_{-\infty}^{+\infty} x(p, \theta) f_{P\Theta}(p, \theta) \, dp \, d\theta$$

$$= \int_0^1 \int_0^{2\pi} p \cos(\omega t + \theta) \frac{1}{2\pi} \, d\theta \, dp$$

$$= \frac{1}{2} \int_0^{2\pi} \cos(\omega t + \theta) \frac{1}{2\pi} \, d\theta = 0.$$

The autocorrelation function is given by

$$R_X(t_1, t_2) = \int_{-\infty}^{+\infty} \int_{-\infty}^{+\infty} x(t_1) x(t_2) f_{P\Theta}(p, \theta) \, d\theta \, dp$$

$$= \int_0^1 \int_0^{2\pi} p^2 \cos(\omega t_1 + \theta) \cos(\omega t_2 + \theta) \frac{1}{2\pi} \, d\theta \, dp$$

$$= \frac{1}{3} \int_0^{2\pi} \cos(\omega t_1 + \theta) \cos(\omega t_2 + \theta) \frac{1}{2\pi} \, d\theta$$

$$= \tfrac{1}{6} \cos(\omega t_1 - \omega t_2) = \tfrac{1}{6} \cos \omega \tau.$$

In the above examples the functional relationship between a random variable and the value of the stochastic process is expressed by an equation. We now consider a slightly different case. In Example 5 the stochastic process is still specified by a transformation of variables, but there is no equation to express this transformation.

Example 4 leads into the more general case considered in Example 5.

4. Each sample function from the stochastic process $X(t, \zeta)$ is a periodic square wave with amplitude A. The starting time t_0 is uniformly distributed $(0, T)$. A typical sample function is shown in Fig. 10.8. Find the mean and autocorrelation functions.

Solution. This process is ergodic in the mean and in autocorrelation (at least). The easy way to solve the problem is to use Eqs. (9.16) and (9.17). However, this will not help us with the next example, which is not ergodic. As a check

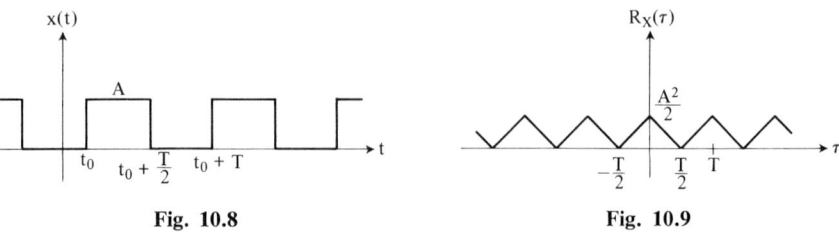

Fig. 10.8 Fig. 10.9

on our solution, Eq. (9.16) for the mean gives

$$m_X(t) = A/2,$$

and Eq. (9.17) for the autocorrelation gives the function plotted in Fig. 10.9.

We now wish to find $f_X(x; t)$ in order to solve for the mean. Consider values of t only in the interval $(t_0 < t < t_0 + T)$. Obviously (see Fig. 10.8),

$$x(t) = \begin{cases} A & \text{if } t_0 < t < (t_0 + T/2), \\ 0 & \text{if } (t_0 + T/2) < t < (t_0 + T). \end{cases}$$

Therefore we have the equivalent events

$$\{X = A\} \sim \{t_0 < t < (t_0 + T/2)\},$$
$$\{X = 0\} \sim \{(t_0 + T/2) < t < (t_0 + T)\}.$$

The probability of each of these events is 0.5. The pdf $f_X(x; t)$ is shown in Fig. 10.10. The mean value is thus

$$m_X(t) = \int_{-\infty}^{+\infty} x f_X(x; t)\, dx = 0 \cdot \tfrac{1}{2} + A \cdot \tfrac{1}{2} = \frac{A}{2}.$$

In order to solve for the autocorrelation function, we must find the joint pdf $f_X(x_1, x_2; t_1, t_2)$. That is, we cannot use Eq. (6.14) in this type of stochastic process, since there is no formula relating the random variable t_0 and the member function $x(t)$. (At least, not in the next example.) Since X is a discrete random variable, the joint pdf is given by

$$f_X(x_1, x_2; t_1, t_2) = P\{X(t_1) = x_1, X(t_2) = x_2\}.$$

There are two possible values of $X(t_1)$, namely, 0 and A, and similarly for $X(t_2)$. Thus there are four possibilities (combinations) of x_1 and x_2.

First consider $x_1 = A$, $x_2 = A$, and compute the pdf

$$f_X(A, A; t_1, t_2) = P\{X(t_1) = A, X(t_2) = A\}.$$

This probability depends on the separation $t_2 - t_1$. Consider Fig. 10.11, and suppose $t_2 - t_1 < T/2$ with $t_1 < t_2$. The event $\{X(t_1) = A, X(t_2) = A\}$ occurs if

$$t_2 < t_0 + (K + \tfrac{1}{2})T \qquad (10.14)$$

10.2 CALCULATING MOMENTS OF STOCHASTIC PROCESSES

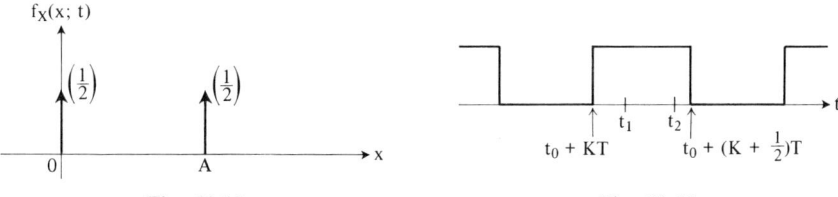

Fig. 10.10   Fig. 10.11

and
$$t_1 > t_0 + KT. \tag{10.15}$$

Combining inequalities (10.14) and (10.15), we have
$$t_2 - (K + \tfrac{1}{2})T < t_0 < t_1 - KT.$$

The probability of this event is
$$P\{t_2 - (K + \tfrac{1}{2})T < t_0 < t_1 - KT\} = \int_{t_2-(K+1/2)T}^{t_1-KT} f_{T_0}(t_0)\,dt_0 = \frac{1}{2} - \frac{t_2 - t_1}{T}$$

or
$$f_X(A, A; t_1, t_2) = \frac{1}{2} - \frac{t_2 - t_1}{T}, \qquad 0 < t_2 - t_1 < \frac{T}{2}.$$

By similar reasoning,
$$f_X(0, 0; t_1, t_2) = \frac{1}{2} - \frac{t_2 - t_1}{T}, \qquad 0 < t_2 - t_1 < \frac{T}{2}.$$

Now consider $X(t_1) = A$, $X(t_2) = 0$, still for $0 < t_2 - t_1 < T/2$. This event occurs if
$$t_1 < t_0 + (K + \tfrac{1}{2})T < t_2$$

(see Fig. 10.12). Therefore
$$P\{X(t_1) = A,\ X(t_2) = 0\} = \int_{t_1-(K+1/2)T}^{t_2-(K+1/2)T} f_{T_0}(t_0)\,dt_0 = \frac{t_2 - t_1}{T}$$

or
$$f_X(A, 0; t_1, t_2) = \frac{t_2 - t_1}{T}, \qquad 0 < t_2 - t_1 < \frac{T}{2}.$$

By similar reasoning,
$$f_X(0, A; t_1, t_2) = \frac{t_2 - t_1}{T}, \qquad 0 < t_2 - t_1 < \frac{T}{2}.$$

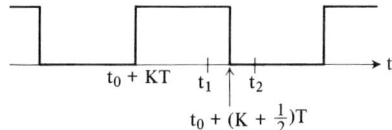

Fig. 10.12

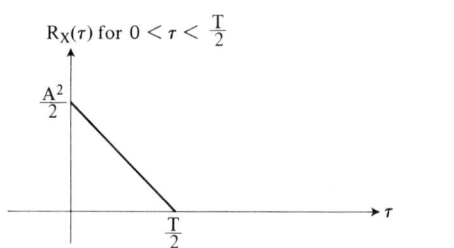

Fig. 10.13

Finally, the autocorrelation function is given by

$$R_X(t_1, t_2) = \int_{-\infty}^{+\infty} \int_{-\infty}^{+\infty} x_1 x_2 f_X(x_1, x_2; t_1, t_2)\, dx_1\, dx_2$$

$$= A^2 \left(\frac{1}{2} - \frac{t_2 - t_1}{T} \right), \quad 0 < t_2 - t_1 < \frac{T}{2}.$$

Substituting $\tau = t_2 - t_1$, we have

$$R_X(\tau) = A^2 \left(\frac{1}{2} - \frac{\tau}{T} \right), \quad 0 < \tau < \frac{T}{2},$$

as shown in Fig. 10.13. Since $R(\tau)$ is an even function in τ, $R_X(\tau) = R_X(-\tau)$. Thus we can plot $R_X(\tau)$ for the interval $-T/2 < \tau < T/2$. Since the stochastic process is periodic, $R_X(\tau)$ is periodic with period T. Thus Fig. 10.9 provides a sketch of this autocorrelation function.

5. Let $X(t, \zeta)$ be a random process with sample functions of the form shown in Fig. 10.14. The value of $x(t)$ in any interval is either a or b, and the value in one interval is statistically independent of the value in any other interval. We could generate such a function, for example, by tossing a coin every T seconds, and allowing the coin to determine whether $x = a$ or $x = b$.

The initial "phase" t_0 is uniformly distributed $(0, T)$:

$$f_{T_0}(t_0) = \frac{1}{T}, \quad 0 < t_0 < T.$$

Also,

$$P\{x = a\} = p \quad \text{and} \quad P\{x = b\} = q.$$

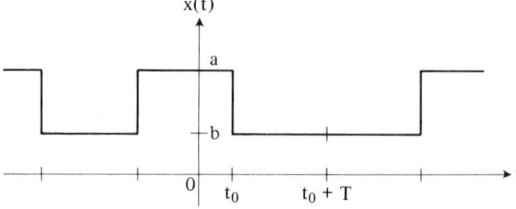

Fig. 10.14

10.2 CALCULATING MOMENTS OF STOCHASTIC PROCESSES

To compute the mean $m_X(t)$, we must find the first-order pdf $f_X(x, t)$. Although we could derive this in terms of f_{T_0}, it is obvious (by inspection) that

$$f_X(x, t) = p\delta(x - a) + q\delta(x - b),$$

which is independent of time. Therefore

$$m_X(t) = ap + bq$$

is also independent of time (is constant).

Finding the second-order pdf is not so simple, for $f_X(x_1, x_2; t_1, t_2)$ is dependent on t_1 and t_2, and the nature of this dependence is not obvious.

Let $0 < t_2 - t_1 < T$. Note that t_1 and t_2 fall in the same interval if (see Fig. 10.15)

$$t_0 + KT < t_1 \tag{10.16}$$

and

$$t_2 < t_0 + (K + 1)T. \tag{10.17}$$

Inequalities (10.16) and (10.17) give

$$t_2 - (K + 1)T < t_0 < t_1 - KT.$$

Let R be the event that t_1 and t_2 fall in the same interval (and then $\bar{R}$ is the event that they do not). Then

$$P(R) = \int_{t_2-(K+1)T}^{t_1-KT} f_{T_0}(t_0)\, dt_0 = 1 - \frac{t_2 - t_1}{T}, \qquad 0 < t_2 - t_1 < T \tag{10.18}$$

and

$$P(\bar{R}) = 1 - P(R) = \frac{t_2 - t_1}{T}, \qquad 0 < t_2 - t_1 < T. \tag{10.19}$$

Recall now that we want to find $f_X(x_1, x_2; t_1, t_2)$. To do this, we will compute $f_X(x_1|x_2, R)$ and $f_X(x_1|x_2, \bar{R})$. Then we can use this to compute $f_X(x_1, x_2|R)$ and $f_X(x_1, x_2|\bar{R})$, so that finally we obtain

$$f_X(x_1, x_2) = f_X(x_1, x_2|R)P(R) + f_X(x_1, x_2|\bar{R})P(\bar{R}). \tag{10.20}$$

Note. For notational simplicity we have dropped the parameters t_1 and t_2 in the discussion leading up to Eq. (10.20), and also in Eq. (10.20). The quantity $f_X(x_1, x_2)$ in Eq. (10.20) certainly does depend on t_1 and t_2, and should be written $f_X(x_1, x_2; t_1, t_2)$.

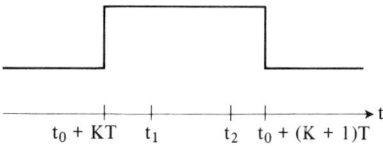

Fig. 10.15

Table 10.1

$f_X(x_1\|x_2, R)$	$x_2 = a$	$x_2 = b$
$x_1 = a$	1	0
$x_1 = b$	0	1

Table 10.2

$f_X(x_1\|x_2, \bar{R})$	$x_2 = a$	$x_2 = b$
$x_1 = a$	p	p
$x_1 = b$	q	q

Table 10.3

$f_\mathbf{X}(x_1, x_2\|R)$	$x_2 = a$	$x_2 = b$
$x_1 = a$	p	0
$x_1 = b$	0	q

There are two possible values for x_1 and two for x_2, namely, a and b. Thus there are four possible combinations. Tables 10.1 through 10.6 are a concise way of displaying the probabilities for these combinations. The entries in these tables are probabilities. For example, in Table 10.2 the entry for $x_1 = a$ and $x_2 = a$ is the probability that $x_1 = a$ given that $x_2 = a$ and also given $\bar{R}$. Since the value of x_1 is statistically independent of the value of x_2 for $\bar{R}$, this probability is p.

The values in Tables 10.3 and 10.4 are calculated from those in Tables 10.1 and 10.2, respectively, by use of formulas such as

$$f_\mathbf{X}(x_1, x_2|\bar{R}) = f_X(x_1|x_2, \bar{R})f_X(x_2).$$

Finally, the entries in Table 10.5 are computed by use of formulas such as

$$f_\mathbf{X}(x_1, x_2) = f_\mathbf{X}(x_1, x_2|R)P(R) + f_\mathbf{X}(x_1, x_2|\bar{R})P(\bar{R}).$$

Keep in mind that Table 10.5 is for $0 < t_2 - t_1 < T$. For $t_2 - t_1 > T$, $X(t_1)$ and $X(t_2)$ cannot both fall in the same interval. Hence $X(t_1)$ and $X(t_2)$ are statistically independent, and the values of $f_\mathbf{X}$ are as shown in Table 10.6.

Table 10.4

$f_X(x_1, x_2 \mid \bar{R})$	$x_2 = a$	$x_2 = b$
$x_1 = a$	p^2	pq
$x_1 = b$	pq	q^2

Table 10.5
Values of $f_X(x_1, x_2; t_1, t_2)$ for $0 < t_2 - t_1 < T$

$f_X(x_1, x_2)$	$x_1 = a,\ x_2 = a$	$x_1 = a,\ x_2 = b$	$x_1 = b,\ x_2 = a$	$x_1 = b,\ x_2 = b$
R	$p \cdot P(R)$	0	0	$q \cdot P(R)$
$\bar{R}$	$p^2 \cdot P(\bar{R})$	$pq \cdot P(\bar{R})$	$pq \cdot P(\bar{R})$	$q^2 \cdot P(\bar{R})$

Table 10.6
Values of $f_X(x_1, x_2; t_1, t_2)$ for $t_2 - t_1 > T$

$f_X(x_1, x_2)$	$x_2 = a$	$x_2 = b$
$x_1 = a$	p^2	pq
$x_1 = b$	pq	q^2

We are now able to compute the autocorrelation function. For $0 < t_2 - t_1 < T$, we have, using Table 10.5,

$$R_X(t_1, t_2) = E[X(t_1)X(t_2)] = a^2[p \cdot P(R) + p^2 \cdot P(\bar{R})]$$
$$+ ab[0 + 2pq \cdot P(\bar{R})] + b^2[q \cdot P(R) + q^2 \cdot P(\bar{R})].$$

Substitution of Eqs. (10.18) and (10.19) gives (after some algebra, with $t_2 - t_1 = \tau$)

$$R_X(\tau) = a^2 p + b^2 q - pq(a-b)^2 \frac{\tau}{T}, \qquad 0 < \tau < T.$$

For $\tau > T$, we have, using Table 10.6,

$$R_X(\tau) = a^2 p^2 + 2abpq + b^2 q^2 = (ap + bq)^2, \qquad \tau > T.$$

Finally, since $R_X(\tau)$ is even in τ, we can find $R_X(-\tau)$ from $R_X(\tau)$. The complete function is plotted in Fig. 10.16.

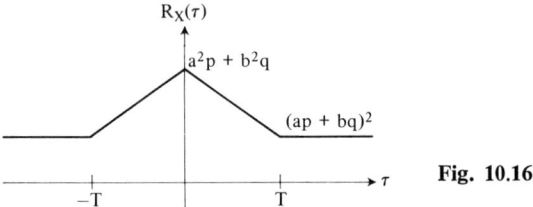

Fig. 10.16

In summary, in this section we have been concerned with calculating the mean and autocorrelation function for a stochastic process that is specified by a transformation of variables. (These same methods apply to the calculation of any order moment.) Two cases were considered: where the relationship is expressed by an equation and where the relationship cannot be expressed by an equation. In the first case we simplified the solution by working in the range space of the random variable (Eq. 6.14). In the second case (Example 5) this was not possible, and we obtained the solution by working in the range space of the stochastic process (Eqs. 9.3 and 9.4). In the next section we will consider these problems in relation to systems.

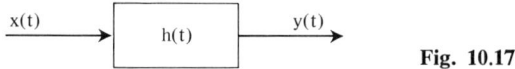

Fig. 10.17

10.3 CALCULATING MOMENTS OF STOCHASTIC PROCESSES SPECIFIED BY OPERATORS (SYSTEMS)

Given a linear system with impulse response $h(t)$ and system function $H(f)$, as shown in Fig. 10.17, we now consider the following two related problems:

a) What must we know about the input x in order to compute certain statistics in the output y?

b) How can we compute these statistics?

Suppose we wish to compute the mean value of the output, $m_Y(t)$. Can we compute this mean from a knowledge of the input mean, $m_X(t)$, or must we know more about the input? Suppose we want to know the cdf $F_Y(y; t)$. Is knowledge of $F_X(x; t)$ sufficient? We are assuming that $h(t)$ is known.

Answers to these and other questions will be attempted in the following discussion.

Computing the Mean of y in the Time Domain

For a linear system the output $y(t)$ is related to the input $x(t)$ by convolution:

$$y(t) = \int_{-\infty}^{+\infty} x(t - \lambda) h(\lambda) \, d\lambda. \tag{10.21}$$

10.3 MOMENTS OF STOCHASTIC PROCESSES SPECIFIED BY OPERATORS

If $x(t)$ is a sample function from a stochastic process, then Eq. (10.21) holds for the entire process $X(t, \zeta)$. In the following discussion, to simplify the equations, we drop the parameter ζ, but dependence on ζ is understood. We then have

$$Y(t) = \int_{-\infty}^{+\infty} X(t - \lambda)h(\lambda)\,d\lambda, \qquad (10.22)$$

and the mean $m_Y(t)$ may be computed from

$$m_Y(t) = E[Y(t)] = E\int_{-\infty}^{+\infty} X(t - \lambda)h(\lambda)\,d\lambda$$

$$= \int_{-\infty}^{+\infty} E[X(t - \lambda)]h(\lambda)\,d\lambda$$

$$= \int_{-\infty}^{+\infty} m_X(t - \lambda)h(\lambda)\,d\lambda. \qquad (10.23)$$

Notes. a) If $X(t, \zeta)$ is stationary in the mean, then so is $Y(t, \zeta)$. The average of X is a constant, $m_X = E[X(t - \lambda)]$, and therefore is independent of the variable of integration in Eq. (10.23). This constant can be factored out of the integral to yield

$$m_Y = m_X \int_{-\infty}^{+\infty} h(\lambda)\,d\lambda. \qquad (10.24)$$

Thus m_Y is the product of m_X and the area under the impulse response.

b) The exchange of expectation and integration operations in Eq. (10.23) is permissible if the following two conditions are satisfied:

$$\int_{-\infty}^{+\infty} E[|X(t - \lambda)|]\,|h(\lambda)|\,d\lambda < \infty$$

and

$$|X(t)| < \infty \qquad \text{for any } t.$$

These conditions are always satisfied in practice for real signals and stable systems.

c) The only knowledge we need about the input signal in order to calculate $m_Y(t)$ is the mean function $m_X(t)$. According to Eq. (10.23), m_Y is the convolution of m_X with the impulse response.

Computing the Mean of y in the Frequency Domain

Suppose the input process $X(t, \zeta)$ is stationary (of order one or higher), so that $m_X(t)$ is constant with time. Therefore the Fourier transform of the mean, $M_X(f)$, is a delta function at zero frequency, since $m_X(t)$ is just a dc-level.

For a nonstationary process (of any order) the mean $m_X(t)$ changes with time. Therefore $M_X(f)$ contains nonzero frequencies. Thus we can find the mean of Y in the frequency domain by taking the Fourier transform of Eq. (10.23), obtaining

$$M_Y(f) = M_X(f)H(f). \tag{10.25}$$

Note. If the input process is stationary in the mean, then in the frequency domain both M_X and M_Y should be delta functions. From Eq. (10.25), with $M_X(f) = m_X \delta(f)$, we have

$$M_Y(f) = m_X \delta(f) H(0).$$

Calculating the Autocorrelation of y in the Time Domain

The autocorrelation function for the output, $R_Y(t_1, t_2)$, is the average value of the product $Y(t_1)Y(t_2)$. First we use Eq. (10.22) to compute $Y(t_1)$ and then $Y(t_2)$. Then we form the product, obtaining

$$Y(t_1)Y(t_2) = \int_{-\infty}^{+\infty} X(t_1 - \lambda)h(\lambda)\,d\lambda \int_{-\infty}^{+\infty} X(t_2 - \gamma)h(\gamma)\,d\gamma$$

$$= \int_{-\infty}^{+\infty}\int_{-\infty}^{+\infty} X(t_1 - \lambda)X(t_2 - \gamma)h(\lambda)h(\gamma)\,d\lambda\,d\gamma.$$

Finally, we take the average value:

$$R_Y(t_1, t_2) = E[Y(t_1)Y(t_2)]$$

$$= E \int_{-\infty}^{+\infty}\int_{-\infty}^{+\infty} X(t_1 - \lambda)X(t_2 - \gamma)h(\lambda)h(\gamma)\,d\lambda\,d\gamma$$

$$= \int_{-\infty}^{+\infty}\int_{-\infty}^{+\infty} E[X(t_1 - \lambda)X(t_2 - \gamma)]h(\lambda)h(\gamma)\,d\lambda\,d\gamma$$

$$= \int_{-\infty}^{+\infty}\int_{-\infty}^{+\infty} R_X(t_1 - \lambda, t_2 - \gamma)h(\lambda)h(\gamma)\,d\lambda\,d\gamma. \tag{10.26}$$

Notes. a) The autocorrelation function R_X is all we need to know about the input to compute the output autocorrelation function R_Y.

b) If $X(t, \zeta)$ is wide-sense stationary, then so is $Y(t, \zeta)$. Replace $R_X(t_1 - \lambda, t_2 - \gamma)$ in Eq. (10.26) by $R_X(t_1 - \lambda - t_2 + \gamma) = R_X(\tau - \lambda + \gamma)$. The output autocorrelation $R_Y(\tau)$ is thus a function of the time difference $\tau = t_1 - t_2$.

c) If $X(t, \zeta)$ is wide-sense stationary, then the mean-square value of the output is $R_Y(0)$:

$$R_Y(0) = \int_{-\infty}^{+\infty}\int_{-\infty}^{+\infty} R_X(\gamma - \lambda)h(\lambda)h(\gamma)\,d\lambda\,d\gamma. \tag{10.27}$$

Note that knowledge of the input mean-square value $R_X(0)$ is not sufficient to compute $R_Y(0)$. The input autocorrelation function must be known.

10.3 MOMENTS OF STOCHASTIC PROCESSES SPECIFIED BY OPERATORS

Calculating the Spectral Density of y in the Frequency Domain

The autocorrelation and power spectral density functions form a Fourier transform pair for (wide-sense) stationary processes. Equation (10.26) enables us to calculate the autocorrelation function for general (nonstationary) processes, but for the special case of stationary processes this same information may be obtained in the frequency domain. From Eq. (2.28) the output power spectral density is given by

$$G_Y(f) = G_X(f)|H(f)|^2, \tag{10.28}$$

and then the autocorrelation function $R_Y(\tau)$ is found by the inverse Fourier transform.

Notes. a) In many cases of interest the desired information about y is the power spectral density function. Hence Eq. (10.28) is sufficient.

b) A gaussian process is specified by knowledge of $m_Y(t)$ and $R_Y(\tau)$. The procedures just discussed then enable us to specify the output process if the system is linear, and if the input process is gaussian.

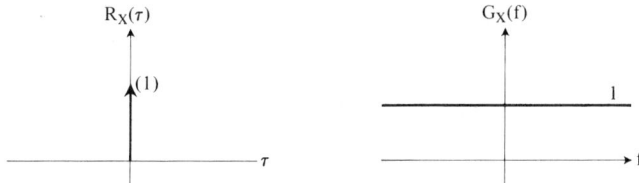

Fig. 10.18 White Noise.

Example 6. The simplest example is for a stationary input with known mean and autocorrelation applied to a low-pass filter. Suppose

$$m_X(t) = 0, \quad R_X(t_1, t_2) = \delta(t_1 - t_2) = \delta(\tau).$$

Thus the input is white noise with zero mean (Fig. 10.18). The term "white noise" comes from an analogy with white light: all frequencies are present. Note that $\delta(\tau) \leftrightarrow 1$, so that $G_X(f)$ is a constant with frequency.

The low-pass filter of Fig. 10.19 has the impulse response $h(t)$ and the transfer function $H(f)$ shown. By Eq. (10.24) the output mean is zero. We will calculate $R_Y(\tau)$ in the time domain and then in the frequency domain.

Time domain. For the output autocorrelation, we set $R_X(t_1 - \lambda, t_2 - \gamma) = R_X(\tau - \lambda + \gamma) = \delta(\tau - \lambda - \gamma)$ in Eq. (10.26):

$$R_Y(\tau) = \int_0^\infty \int_0^\infty \delta(\tau - \lambda + \gamma) a e^{-a\lambda} a e^{-a\gamma} \, d\lambda \, d\gamma.$$

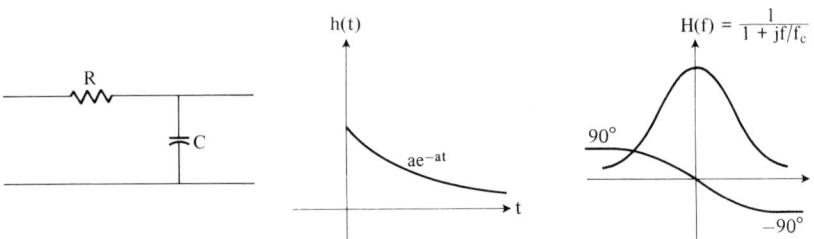

Fig. 10.19 Low-pass RC-filter $a = 1/RC = \omega_c$.

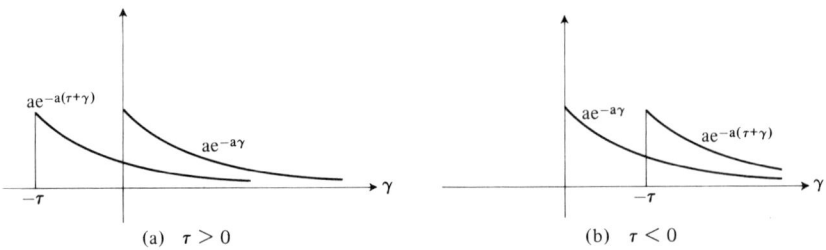

Fig. 10.20

The lower limit on the integrals is zero because the impulse response is zero for $t < 0$.

The first integration (with respect to λ) yields a function of τ and γ. Call it $I(\tau, \gamma)$. Now this integration is the convolution of $h(\tau)$ with $\delta(\tau + \gamma)$. The result is

$$I(\tau, \gamma) = \int_0^\infty \delta(\tau + \gamma - \lambda) a e^{-a\lambda}\, d\lambda$$
$$= \begin{cases} 0, & \tau + \gamma < 0 \text{ or } \tau < -\gamma, \\ ae^{-a(\tau+\gamma)}, & \tau + \gamma > 0 \text{ or } \tau > -\gamma. \end{cases}$$

To find $R_Y(\tau)$, we multiply $I(\tau, \gamma)$ by $ae^{-a\gamma}$ and integrate. Care must be taken, however, for the value of τ affects the value of the integral. The two functions $I(\tau, \gamma)$ and $h(\gamma)$ are shown in Fig. 10.20, first for $\tau > 0$ and then for $\tau < 0$. For $\tau > 0$,

$$R_Y(\tau) = a^2 e^{-a\tau} \int_0^\infty e^{-2a\gamma}\, d\gamma = \frac{a}{2} e^{-a\tau}.$$

Since $R_Y(\tau)$ is an even function of τ, we can find $R_Y(\tau)$ for $\tau < 0$ from $R_Y(\tau)$ for $\tau > 0$. As a check, for $\tau < 0$,

$$R_Y(\tau) = a^2 e^{-a\tau} \int_{-\tau}^\infty e^{-2a\gamma}\, d\gamma = \frac{a}{2} e^{a\tau},$$

which is the correct answer. $R_Y(\tau)$ is plotted in Fig. 10.21.

10.3 MOMENTS OF STOCHASTIC PROCESSES SPECIFIED BY OPERATORS

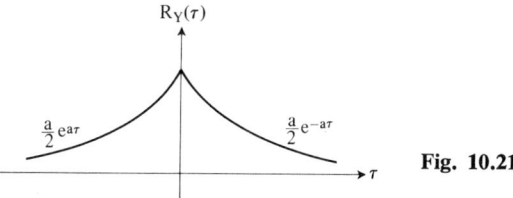

Fig. 10.21

Frequency domain. Here we will find the spectral density $G_Y(f)$ using Eq. (10.28). We have

$$|H(f)|^2 = H(f)H^*(f) = \frac{1}{1 + \left(\frac{\omega}{\omega_c}\right)^2},$$

so that

$$G_Y(f) = 1 \frac{1}{1 + \left(\frac{\omega}{\omega_c}\right)^2} = \frac{a^2}{a^2 + \omega^2}, \quad -\infty < f < +\infty,$$

which is the Fourier transform of $R_Y(\tau)$ shown in Fig. 10.21.

Notes. a) In this case the frequency-domain approach was considerably shorter. This is often the case, but when $R_Y(\tau)$ is desired, it can sometimes be difficult to transform the resulting $G_Y(f)$.

b) The frequency-domain approach is useful only if the input process is stationary. Otherwise it has no meaning.

Calculating the Distribution of y

Given the input cdf $F_X(x; t) = P\{X(t) \leq x\}$, let us determine what is needed to calculate the cdf of the output of a linear system. We are restricting this discussion to linear systems so that the convolution integral can be used. The cdf of the output $Y(t)$ is

$$F_Y(y; t) = P\{Y(t) \leq y\} = P\left\{\int_{-\infty}^{+\infty} X(t - \lambda)h(\lambda)\,d\lambda \leq y\right\}.$$

For a particular value of t, the integral $\int_{-\infty}^{+\infty} X(t - \lambda)h(\lambda)\,d\lambda$ is a random variable. We wish to compute the probability that this random variable Y will be less than or equal to y. The experimental outcome ζ determines whether or not $Y \leq y$, and so in an attempt to solve this problem the event $\{Y \leq y\}$ should be related to the sample space S by equivalent events. Unfortunately, it is generally difficult (or impossible) to find these equivalent events.

The difficulty in this problem arises because of system memory. Figure 10.22 should help explain what is meant by memory. Two system impulse responses, $h_1(t)$ and $h_2(t)$, are shown. The response $h_1(t)$ has no memory, and

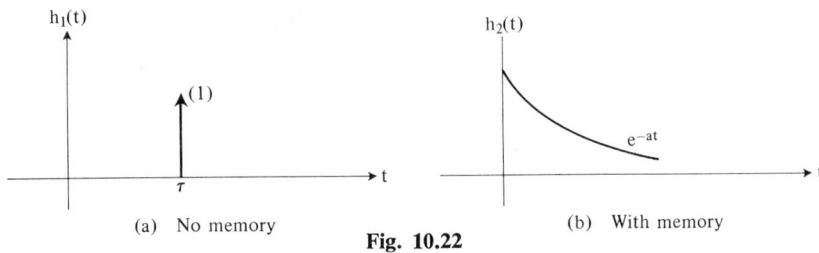

(a) No memory (b) With memory

Fig. 10.22

$h_2(t)$ does have memory. The first impulse response, $h_1(t)$, represents an ideal delay line in which the input is simply delayed by a time τ. The second, $h_2(t)$, is the impulse response of a low-pass filter, for example, the RC-circuit of Fig. 10.19. Another system with no memory is shown in Fig. 10.23(a). The simple voltage divider has the impulse response shown in Fig. 10.23(b). In general, systems with energy-storage elements have memory and simple voltage-divider circuits do not. (However, there are exceptions to this rule.)

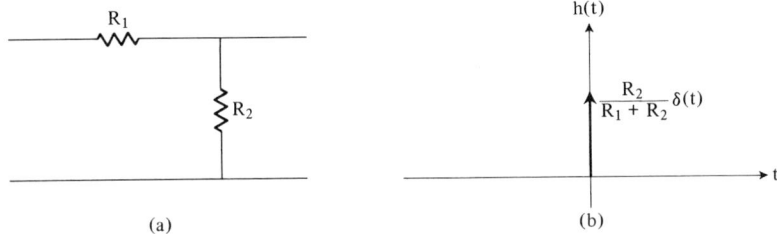

Fig. 10.23. Simple voltage divider and impulse response.

In Chapter 7 we discussed the problem of finding $F_Y(y; t)$ for systems with no memory. Unfortunately, the problem is difficult for systems with memory, and no general solution exists. However, the problem has been solved for special cases.

An obvious example of a case in which a solution exists is when the input is gaussian. The pdf (and cdf) in this case is determined if the first two moments are known. Thus m_Y and R_Y, the mean and autocorrelation functions of the output, are sufficient to determine the pdf of the output. This is one example of the analytical simplicity that results when the gaussian process is used.

Specific solutions have been found for a few other cases, but all of these are special cases and no general procedure exists. Therefore we will leave this difficult problem and in the next chapter turn our attention to mean-square estimation.

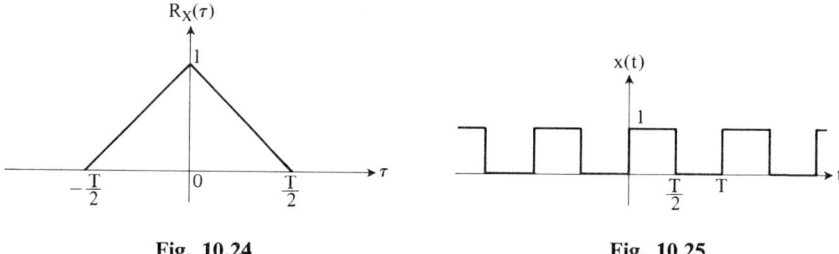

Fig. 10.24   Fig. 10.25

PROBLEMS

1. Let X be a wide-sense stationary random process with autocorrelation $R_X(\tau)$, shown in Fig. 10.24. Find and plot the power spectral density.
2. Let X have a periodic autocorrelation function with period T. One period of this function is shown in Fig. 10.24. Find and plot the power spectral density.
3. Let X be a periodic ergodic random process. One sample function is shown in Fig. 10.25.
a) Find and plot the autocorrelation and power spectral density.
b) Must the process be ergodic before you can solve the problem? Why?
4. A sample function from a stochastic process is given by

$$X(t) = At, \quad -\infty < t < +\infty,$$

where A is a random variable with normal distribution (0, 1). Find
a) the mean $m_X(t)$,
b) the autocorrelation function $R_X(t_1, t_2)$.
5. A sample function from a stochastic process is given by

$$X(t) = \cos \omega t,$$

where ω is a random variable uniformly distributed between ω_1 and ω_2. Find
a) the mean $m_X(t)$,
b) the autocorrelation function $R_X(t_1, t_2)$.
6. A typical function from a random process $Y(t)$ is shown in Fig. 10.26. The phase t_0 is uniformly distributed $(0, T)$, and the amplitude y_i is gaussian $(0,1)$.

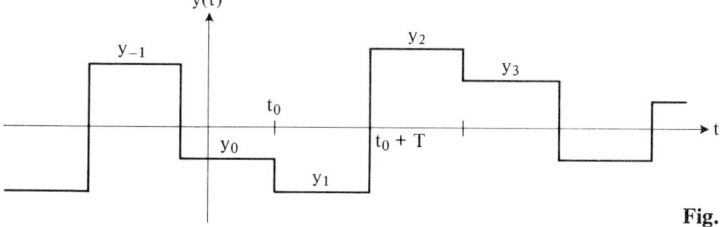

Fig. 10.26

The amplitudes in different intervals are statistically independent. Find
a) the mean $m_Y(t)$,
b) the autocorrelation function $R_Y(t_1, t_2)$.

7. The power spectral density for a random process is

$$G_X(\omega) = \frac{\omega^2}{\omega^6 + 1}.$$

Find
a) the mean-square value,
b) the autocorrelation function of the process.

8. White noise with spectral density η is applied to a low-pass RC-filter. The product RC equals 1. Find
a) the power spectral density of the output,
b) the autocorrelation function of the output.

9. Let

$$y(t) = \frac{dx(t)}{dt}.$$

Given that $X(t)$ is a stochastic process with spectral density $G_X(f)$, find $G_Y(f)$ in terms of $G_X(f)$. [*Hint*: Use the differentiation property of Fourier transforms.]

10. Repeat Problem 9 for

$$y(t) = \int_{-\infty}^{t} x(\lambda)\, d\lambda.$$

FURTHER READING

1. A. PAPOULIS, *Probability, Random Variables, and Stochastic Processes*, McGraw-Hill, New York, 1965.
2. J. M. WOZENCRAFT and I. M. JACOBS, *Principles of Communication Engineering*, Wiley, New York, 1965.

Both of these texts provide a good discussion of all the material in this chapter.

11
MEAN-SQUARE ESTIMATION

11.1 INTRODUCTION

During the early development of radar in (and before) the second world war, Norbert Wiener at M.I.T. solved the famous prediction problem. The problem is this: Given the past flight path of an approaching enemy aircraft (as recorded by radar), where does one aim the gun in order to shoot down the aircraft? That is, how does one use the data about the past positions of the aircraft to predict its position at some future time?

This is only one of a class of problems solved by Wiener at the time. His general approach to this problem was applicable to other problems, also. For example, his approach could be used to determine how to separate signal from noise in a communications system. Since that time the problem and solution have been generalized, and now Wiener's original work is but a special case of the more general theory.

To understand this problem, one must realize that there is no clear-cut solution. In most scientific problems the solution is either correct or incorrect, and no personal judgment is involved. In the prediction problem one is faced with the task of making the "best" estimate, and judgment must be used in deciding what is "best." Obviously, the best estimate is the one that gives the exact value of the quantity being estimated, but, due to unpredictable fluctuations, some error is unavoidable.

There are several criteria we could use to find the best solution to the prediction problem. For example, we might choose to maximize the probability of hitting the aircraft (maximum likelihood estimation). Certainly this is a reasonable criterion but we encounter mathematical difficulties. Another reasonable criterion is to minimize the mean-square error, and this is the criterion used by Wiener. There are other criteria that could be used, but all lead to mathematical difficulties. For this reason we will discuss only mean-square estimation.

11.2 MEAN-SQUARE ESTIMATION

Minimizing the mean-square error is a useful criterion in many applications because it leads directly and simply to a solution. Many other (perhaps better) criteria are not so easy to use because of the mathematical difficulties involved. The *error* is the difference between the outcome and our estimate (guess) of the outcome:

$$e = X - \hat{X}.$$

The outcome X is a random variable. Our estimate $\hat{X}$ is a number. Hence the error e is a random variable. This quantity is squared and then averaged to obtain

$$\overline{e^2} = E[(X - \hat{X})^2]. \tag{11.1}$$

The value of $\hat{X}$ that minimizes this expression is the minimum mean-square estimate of X.

In the following, we will discuss several related, progressively more complex estimation problems leading up to the filtering and prediction problem.

Estimating the Random Variable X by a Constant

Suppose we know the distribution of X and want to estimate the value of the random variable $X(\zeta)$ by a constant a. The error e equals $X - a$, and the mean-square error is

$$\overline{e^2} = E[(X - a)^2]$$
$$= \int_{-\infty}^{+\infty} (x - a)^2 f_X(x)\, dx$$
$$= a^2 - 2aE(X) + E(X^2).$$

Differentiating with respect to a and setting the derivative equal to zero, we have

$$2a - 2E(X) = 0 \quad \text{or} \quad a = E(X).$$

Therefore we should use the average $E(X)$ to estimate X when we wish to minimize the mean-square error.

Notes. a) The experiment is performed with outcome ζ. This outcome ζ then determines a value of the random variable X, and it is this number that we estimate.

b) In this problem the estimate is made before the experiment is performed (or at least before we have any information about the experimental outcome). In all the following we will assume that some information is available about the experimental outcome, and if this information can be incorporated into our estimate, it should improve our accuracy.

c) Note how the criterion of minimizing the mean-square error led directly to the solution.

11.2 MEAN-SQUARE ESTIMATION 225

Estimating the Random Variable X given $Y(\zeta)$

Suppose two random variables X and Y are defined on the same sample space. The experiment is performed, but the outcome ζ is unknown. However, we are allowed to observe the value $Y(\zeta)$, and the problem is as follows: We know $F_X(x)$ and $Y(\zeta)$ and want to estimate the value of $X(\zeta)$ by a function of Y, $g(Y)$. Here the estimate X is a function of Y, $\hat{X} = g(Y)$, so the mean-square error is

$$\overline{e^2} = E\{[X - g(Y)]^2\}.$$

The function g that minimizes this expression is

$$g(Y) = E(X|Y),$$

as shown below:

$$E\{[X - g(Y)]^2\} = \int_{-\infty}^{+\infty}\int_{-\infty}^{+\infty} [x - g(y)]^2 f_{XY}(x, y)\, dx\, dy$$

$$= \int_{-\infty}^{+\infty}\int_{-\infty}^{+\infty} [x - g(y)]^2 f_X(x|y) f_Y(y)\, dx\, dy$$

$$= \int_{-\infty}^{+\infty} f_Y(y) \int_{-\infty}^{+\infty} [x - g(y)]^2 f_X(x|y)\, dx\, dy. \quad (11.2)$$

Let

$$h(y) = \int_{-\infty}^{+\infty} [x - g(y)]^2 f_X(x|y)\, dx, \quad (11.3)$$

so that Eq. (11.2) can be written as

$$E\{[X - g(Y)]^2\} = \int_{-\infty}^{+\infty} f_Y(y) h(y)\, dy.$$

Now consider Fig. 11.1. We wish to minimize the area under the product $f_Y(y)h(y)$. We have no control over $f_Y(y)$; this is the marginal pdf of the random variable Y. Hence we can operate only on $h(y)$. If the function $g(y)$ in Eq. (11.3) is chosen so as to minimize $h(y)$ for every value of y, then this will minimize the mean-square error.

Expand Eq. (11.3) to obtain

$$h(y) = g^2(y) - 2g(y)E(X|y) + E(X^2|y).$$

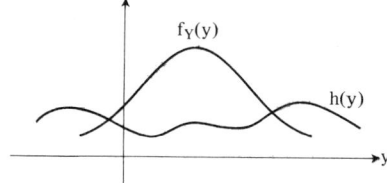

Fig. 11.1

Setting the derivative equal to zero gives

$$\frac{dh(y)}{dy} = 2g(y) - 2E(X|y) = 0$$

or

$$g(y) = E(X|y). \quad (11.4)$$

Notes. a) This situation differs from the first (estimating X by a constant) in that the experiment is performed with outcome ζ, and we are allowed to observe the value of $Y(\zeta)$. If knowledge of $Y(\zeta)$ allows us to determine ζ, then we can estimate $X(\zeta)$ exactly. That is, $E(X|y) = X$.

Of course, this is the trivial case in which knowledge of Y gives us complete information about X. The other extreme is when knowledge of Y gives us no information about X, and $E(X|y) = E(X)$.

b) The function $g(y)$ determined by Eq. (11.4) is, in general, not a straight line. The problem of finding an expression for $g(y)$ in the general case is greatly simplified if $g(y)$ is restricted to be a straight line. This is the linear mean-square estimation problem, which we will consider after some examples.

c) In the above discussion we made the statement that minimizing $h(y)$ for every value of y will minimize the integral

$$\int_{-\infty}^{+\infty} f_Y(y)h(y)\,dy.$$

This statement is true since both f_Y and h are positive for every value of y. As a counterexample, suppose f_Y is negative for values of y in some interval, and positive elsewhere. Then any $h(y)$ can be chosen to make the integral zero, so long as the product $f_Y(y)h(y)$ has just as much negative area as positive area.

Table 11.1

y	4	1	0	1	4	9
x	-2	-1	0	1	2	3
ζ	•	••	••	••	••	•••

Examples

The random variables X and Y are defined in Table 11.1. The experiment consists of rolling a single die, and the experimental outcome determines a value for X and Y. We now consider several examples to illustrate the above concepts.

1. Find a constant a that will give a minimum mean-square estimate of X and calculate the resulting mean-square error.

11.2 MEAN-SQUARE ESTIMATION

Solution. We find the constant a to be

$$a = E(X) = 0.5.$$

The resulting error is

$$\overline{e^2} = E[(X - a)^2] = \frac{8.75}{3} \approx 2.92.$$

Note. This resulting error is the average square error. Suppose the value of X is 0. Then the square error is only 0.25. However, if X is 3, then the square error is $(3 - 0.5)^2 = 6.25$. The value 2.92 represents an average of all these square-error values.

2. Given that $Y = 4$, find the minimum mean-square estimate for X and the resulting mean-square error.

Solution. X must be either -2 or $+2$ with equal probability. Hence,

$$\hat{X} = E(X|Y = 4) = 0.$$

The resulting mean-square error is

$$\overline{e^2} = E[(X - \hat{X})^2] = \int_{-\infty}^{+\infty} x^2 f_X(x|y)\, dx$$

$$= (-2)^2 \cdot \tfrac{1}{2} + (2)^2 \cdot \tfrac{1}{2} = 4.$$

Note. The disadvantage of the mean-square error criterion can be seen from this example. Using the criterion of minimizing the mean-square error, we can never guess the correct value of X. Yet if we use $\hat{X} = 2$ as our estimate, the mean-square error will be larger than 4 (Calculate it!), although we have a good chance of being correct.

3. Extend the result of Example 2 for $Y = 0, 1, 9$. This will allow us to plot the "regression curve" $g(Y) = E(X|Y)$.

Solution

$$E(X|Y = 0) = 0,$$
$$E(X|Y = 1) = 0,$$
$$E(X|Y = 9) = 3.$$

Of course, a "curve" cannot be plotted, since the values of $g(y)$ are defined only for $y = 0, 1, 4,$ or 9. Nevertheless, it can be seen that a straight line cannot be drawn through the points on the graph in Fig. 11.2.

The resulting mean-square error is

$$\overline{e^2} = E\{[X - g(Y)]^2\} \approx 1.67.$$

Note. The points on the graph in Fig. 11.2 represent the (nonlinear) mean-square estimates of X for the given values of Y. The "best" approximation

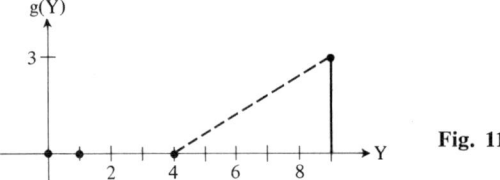

Fig. 11.2

to these points by points on a straight line is the *linear* mean-square estimate. This "best" linear estimate is the straight line that minimizes the mean-square error.

This resulting *linear* mean-square error will usually be larger than the mean-square error (it can never be less). We will compute this linear error in the next example and compare with the above mean-square error.

Linear Estimation of X given Y(ζ)

Suppose we know $F_X(x)$ and $Y(\zeta)$ and want to estimate the value of $X(\zeta)$ by a *linear* function Y,

$$\hat{X} = g(Y) = aY + b. \quad (11.5)$$

Then the mean-square error is

$$\overline{e^2} = E\{[X - (aY + b)]^2\}. \quad (11.6)$$

Note. The only difference between this problem and the previous one is that we are restricting the form of $g(Y)$. It must be a straight line. This greatly simplifies the problem, for now all we have to do is solve for two constants, a and b.

Solution. Again the minimum mean-square error criterion leads directly to a solution. Expanding Eq. (11.6), we have

$$\overline{e^2} = \overline{X^2} + a^2\overline{Y^2} + b^2 - 2a\overline{XY} - 2b\overline{X} + 2ab\overline{Y}.$$

Setting the partial derivative with respect to b equal to zero gives

$$\frac{\partial \overline{e^2}}{\partial b} = 2b - 2\overline{X} + 2a\overline{Y} = 0.$$

The value of b that minimizes Eq. (11.6) is therefore

$$b = m_X - am_Y. \quad (11.7)$$

Using this value of b, we can write Eq. (11.6) as

$$\overline{e^2} = E\{[(X - m_X) - a(Y - m_Y)]^2\}$$
$$= \sigma_X^2 - 2a\mu_{11} + a^2\sigma_Y^2.$$

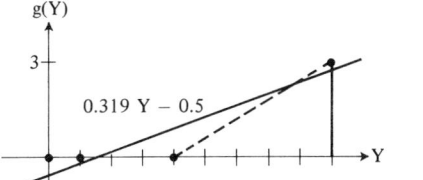

Fig. 11.3

Solving for a by setting the partial derivative with respect to a equal to zero yields

$$a = \frac{\mu_{11}}{\sigma_Y^2}. \tag{11.8}$$

4. Find the linear mean-square estimate of X given Y using the random variables X and Y as defined in the previous examples, and compute the resulting linear mean-square error.

Solution

$$a = \frac{\mu_{11}}{\sigma_Y^2} = \frac{2.91}{9.13} = 0.319,$$

$$b = m_X - a m_Y = 0.5 - (0.319)(3.17) = -0.5,$$

so that $g(Y) = 0.319Y - 0.5$. This function is plotted in Fig. 11.3, with the points from Fig. 11.2 shown for comparison.

The resulting linear mean-square error is

$$\overline{e^2} = E[(X - aY - b)^2] = 1.99.$$

Note that this error compares rather well with the (nonlinear) mean-square error of 1.67 from Example 3. That is, the reduction in complexity resulted in only a slight increase in the mean-square error. This reduction in complexity can be significant in some problems (although not in this example).

11.3 VECTORS AND LINEAR MEAN-SQUARE ESTIMATION

The following example will form the basis for our introduction of vectors in linear mean-square estimation.

Example 5. Let X be a random variable that assumes any of the values $-2, -1, 0, 1, 2$ with equal probability. Let $Y = X + N$, where N is another random variable, statistically independent from X, that assumes any of the values $-1, 0, 1$ with equal probability.

The experiment is performed as follows: A number is selected for X and recorded. A number N_1 is chosen and added to X to form $Y_1 = X + N_1$. A second number N_2, independent of N_1, is chosen and added to the same X to form $Y_2 = X + N_2$. You are shown the two numbers Y_1 and Y_2 and asked to estimate X.

Note. Both X and N have zero mean. We will discuss this situation first, and then extend these concepts to random variables with nonzero mean.

Using the minimum mean-square error criterion, you would choose as the estimate of X that number that minimizes

$$\overline{e^2} = E[(X - \hat{X})^2],$$

where $\hat{X}$ is some function of Y_1, Y_2:

$$\hat{X} = f(Y_1, Y_2). \tag{11.9}$$

The nonlinear estimate is, from our previous discussion,

$$\hat{X} = E(X|Y_1, Y_2).$$

However, our primary interest in this discussion is in the linear mean-square estimate

$$\hat{X} = a_1 Y_1 + a_2 Y_2 = \sum_{i=1}^{2} a_i Y_i. \tag{11.10}$$

For this criterion we will choose the constants a_1 and a_2 so that the mean-square error is minimum.

Now consider two pertinent facts about this example. First, the estimate in Eq. (11.9) is an operator. The domain consists of the random variables Y_1 and Y_2, and the range consists of the random variables $X = \hat{X}$. Second, this operator (transformation) is linear if Eq. (11.10) is used to define the estimate. Therefore the concept of linear transformations on linear vector spaces is applicable to this problem.

A linear vector space was defined in Section 2.2. Random variables satisfy the seven properties of vectors listed there, hence random variables are vectors. Also, the inner product on this space is defined as the expected value,

$$\langle X, Y \rangle = E(XY),$$

so the random variables form an inner product space. Geometric vectors have these same properties and can therefore be used to visualize operations on random variables.

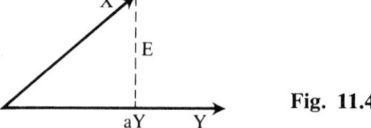

Fig. 11.4

Geometric Vectors and Mean-Square Estimation

Consider the two geometric vectors X and Y shown in Fig. 11.4, and suppose that we wish to estimate X by a component in the direction of Y. The error E

11.3 VECTORS AND LINEAR MEAN-SQUARE ESTIMATION

is a minimum if $|E|^2 = \langle E, E \rangle$ is a minimum. Thus in estimating X we will minimize the squared error functional $|E|^2$.

From Fig. 11.4 the error is given by

$$E = X - aY.$$

We wish to choose a so that $\langle E, E \rangle$ is a minimum:

$$\langle E, E \rangle = \langle (X - aY), (X - aY) \rangle$$
$$= \langle X, X \rangle - 2a \langle X, Y \rangle + a^2 \langle Y, Y \rangle. \quad (11.11)$$

Setting the derivative with respect to a equal to zero yields

$$-2\langle X, Y \rangle + 2a\langle Y, Y \rangle = 0$$

or

$$a = \frac{\langle X, Y \rangle}{\langle Y, Y \rangle}. \quad (11.12)$$

Now note an important feature of this estimation problem. The value of a that minimizes the squared error functional is such that the "data" Y is orthogonal to the error $E = X - aY$; that is,

$$\langle (X - aY), Y \rangle = 0$$

or

$$\langle X, Y \rangle - a \langle Y, Y \rangle = 0, \quad (11.13)$$

which again yields Eq. (11.12). This is easily visualized geometrically from Fig. 11.4.

We will show that this assertion is true in general (for random variables as well as for geometric vectors) by contradiction. Let $\langle (X - aY), Y \rangle = 0$, and suppose that $b \neq a$ is the estimate that minimizes $\langle E, E \rangle$. Then

$$\langle E, E \rangle = \langle (X - bY), (X - bY) \rangle$$
$$= \langle (X - aY + (a - b)Y), (X - aY + (a - b)Y) \rangle$$
$$= \langle (X - aY), (X - aY) \rangle + 2(a - b)\langle (X - aY), Y \rangle$$
$$+ (a - b)^2 \langle Y, Y \rangle.$$

The second term is zero by assumption, and the third term is nonnegative. Therefore

$$\langle (X - bY), (X - bY) \rangle \geq \langle (X - aY), (X - aY) \rangle,$$

which contradicts our assumption that b minimizes the squared error.

Also, the minimum squared error can be found easily by

$$\langle E, E \rangle_{\min} = \langle (X - aY), X \rangle. \quad (11.14)$$

To see this, substitute Eq. (11.12) into (11.11) to obtain

$$\langle E, E \rangle_{\min} = \langle X, X \rangle - \frac{\langle X, Y \rangle^2}{\langle Y, Y \rangle}$$

$$= \langle X, X \rangle - a \langle X, Y \rangle$$

$$= \langle (X - aY), X \rangle.$$

Note. Two important properties of mean-square estimation are illustrated above: (1) the value of a that minimizes the mean-square error is such that the error is orthogonal to the data, and (2) the mean-square error is the inner product of the error and the quantity to be estimated, as shown in Eq. (11.14).

Random Variables and Linear Mean-Square Estimation

Let us assume that the mean values of the random variables X and Y are zero. This will simplify the following discussion, and will enable us to make a direct analogy to geometric vectors. We will then indicate the procedures for random variables with nonzero mean.

Suppose we wish to estimate the random variable X by a linear combination of two or more random variables Y (as in Example 5). The procedures for geometric vectors are still applicable. Suppose we are given n observations of the random variable Y which is statistically related to X. Then the linear mean-square estimate $\hat{X}$ is found by a generalization of Eq. (11.13).

With

$$\hat{X} = \sum_{i=1}^{n} a_i Y_i,$$

the numbers a_i that minimize the mean-square error are such that the error is orthogonal to the data. The error is

$$e = X - \sum_{i=1}^{n} a_i Y_i,$$

so that

$$E\left[\left(X - \sum_{i=1}^{n} a_i Y_i\right) Y_j\right] = 0 \quad \text{for each} \quad j = 1, 2, \ldots, n. \quad (11.15)$$

And the resulting mean-square error is

$$\overline{e^2_{\min}} = E\left[\left(X - \sum_{i=1}^{n} a_i Y_i\right) X\right]. \quad (11.16)$$

For a proof of these more general assertions, see Papoulis [1], Section 11.2.

Solution to Example 5. From Eq. (11.15) we have

$$E[(X - a_1 Y_1 - a_2 Y_2) Y_1] = 0,$$

$$E[(X - a_1 Y_1 - a_2 Y_2) Y_2] = 0,$$

11.3 VECTORS AND LINEAR MEAN-SQUARE ESTIMATION

or two linear equations with two unknowns. The steps are as follows:

$$\overline{XY_1} - a_1\overline{Y_1^2} - a_2\overline{Y_1Y_2} = 0,$$

$$\overline{XY_2} - a_1\overline{Y_1Y_2} - a_2\overline{Y_2^2} = 0,$$

or

$$2 - \tfrac{8}{3}a_1 - 2a_2 = 0,$$

$$2 - 2a_1 - \tfrac{8}{3}a_2 = 0,$$

from which we obtain

$$a_1 = a_2 = \tfrac{3}{7}.$$

From Eq. (11.16) the resulting mean-square error is

$$\overline{e^2} = \overline{X^2} - a_1\overline{Y_1X} - a_2\overline{Y_2X} = \tfrac{2}{7}.$$

Note. If we are given one observation, Y_1, the linear estimate is $\hat{X} = \tfrac{3}{4}Y_1$ and the resulting mean-square error is $e^2 = \tfrac{1}{2}$.

If we are given three observations, Y_1, Y_2, and Y_3, the linear estimate is $\hat{X} = \tfrac{3}{10}Y_1 + \tfrac{3}{10}Y_2 + \tfrac{3}{10}Y_3$, and the resulting mean-square error is $\overline{e^2} = 0.2$.

Generalization to Random Variables with Nonzero Mean

Recall that for a single observation with nonzero mean the best linear estimate was, from Eq. (11.5),

$$\hat{X} = aY + b,$$

where, from Eq. (11.7), $b = m_X - am_Y$. Substitution of this value of b into Eq. (11.5) enables us to write

$$\hat{X} = a(Y - m_Y) + m_X.$$

Now for two observations, Y_1 and Y_2, the best linear estimate is

$$\hat{X} = a_1(Y_1 - m_{Y_1}) + a_2(Y_2 - m_{Y_2}) + m_X.$$

For n observations,

$$\hat{X} = \sum_{i=1}^{n} a_i(Y_i - m_{Y_i}) + m_X. \tag{11.17}$$

As in Eq. (11.15), the coefficients a_i that minimize the linear mean-square error are such that the error is orthogonal to the data. With the estimate X given by Eq. (11.17), the error is

$$X - \hat{X} = X - \sum_{i=1}^{n} a_i(Y_i - m_{Y_i}) - m_X,$$

so that the a_i's satisfy the n linear equations

$$E\left[\left(X - \sum_{i=1}^{n} a_i(Y_i - m_{Y_i}) - m_X\right)Y_j\right] = 0, \quad j = 1, 2, \ldots, n. \tag{11.18}$$

234 MEAN-SQUARE ESTIMATION

As in Eq. (11.16), the resulting mean-square error is the inner product of the error and X, the quantity to be estimated:

$$\overline{e^2} = E\left[\left(X - \sum_{i=1}^{n} a_i(Y_i - m_{Y_i}) - m_X\right)X\right]. \tag{11.19}$$

Example 6. Let us return to Table 11.1 and compute the linear mean-square estimate for X given Y (as in Example 4). Use of Eq. (11.18) yields

$$E[(X - a(Y - m_Y) - m_X)Y] = 0$$

or

$$\overline{XY} - a(\overline{Y^2} - m_Y^2) - m_X m_Y = 0,$$

$$\tfrac{27}{6} - a[\tfrac{115}{6} - (\tfrac{19}{6})^2] - \tfrac{1}{2}(\tfrac{19}{6}) = 0,$$

which yields $a = 0.319$, as before.

Use of Eq. (11.19) for the mean-square error yields

$$\overline{e^2} = \overline{X^2} - a(\overline{XY} - m_Y m_X) - m_X^2$$

$$= \tfrac{19}{6} - (0.319)(\tfrac{27}{6} - \tfrac{19}{12}) - \tfrac{1}{4} = 1.99,$$

again in agreement with the results of Example 4.

11.4 CONTINUOUS DATA

In the preceding sections we discussed the problem of estimating the random variable X when the given data is a finite set of observations $\{Y_i\}$ that are statistically related to X. Now we remove the restriction that the set of observations $\{Y_i\}$ be finite.

For example, consider the prediction problem. The data is the entire past history of the enemy aircraft, say $Y(\lambda)$, $-\infty < \lambda < t$. Thus we are presented with an infinite number of observations (one for each value of λ) and are asked to estimate the value of Y at some time in the future, say $Y(t + \epsilon)$. We could form a linear estimate by weighting each observation $Y(\lambda)$ by a number, say $a(\lambda)$, and summing (integrating). The linear mean-square estimation problem then becomes that of finding the appropriate weights $a(\lambda)$. Using $\hat{Y}(t + \epsilon)$ as the estimate,

$$\hat{Y}(t + \epsilon) = \int_{-\infty}^{t} a(\lambda) Y(\lambda) \, d\lambda,$$

we seek the $a(\lambda)$ that minimizes the mean-square error,

$$\overline{e^2} = E\left\{\left[Y(t + \epsilon) - \int_{-\infty}^{t} a(\lambda) Y(\lambda) \, d\lambda\right]^2\right\}.$$

As another example, consider the filtering problem. The data $Y(\lambda)$ is the sum of signal $X(\lambda)$ plus noise $N(\lambda)$:

$$Y(\lambda) = X(\lambda) + N(\lambda), \quad -\infty < \lambda < t.$$

We wish to estimate the present value of signal $X(t)$. If the estimate $\hat{X}(t)$ is a linear combination of the observations,

$$\hat{X}(t) = \int_{-\infty}^{t} a(\lambda) Y(\lambda) \, d\lambda,$$

then again we seek appropriate weights $a(\lambda)$ that minimize the mean-square error,

$$\overline{e^2} = E\left\{ \left[X(t) - \int_{-\infty}^{t} a(\lambda) Y(\lambda) \, d\lambda \right]^2 \right\}.$$

Note. In this discussion we assume the random variables have zero mean. If they do not, we must make appropriate modifications, as in Section 11.3.

Solution of the Prediction Problem

Applications of the principles introduced in Section 11.3 imply that we should choose those weights $a(\lambda)$ that make the error orthogonal to the data. The solution to the prediction problem is therefore found as follows:

$$E\left\{ \left[Y(t + \epsilon) - \int_{-\infty}^{t} a(\lambda) Y(\lambda) \, d\lambda \right] Y(\gamma) \right\} = 0, \quad -\infty < \gamma < t,$$

or

$$E[Y(t + \epsilon) Y(\gamma)] = \int_{-\infty}^{t} a(\lambda) E[Y(\lambda) Y(\gamma)] \, d\lambda, \quad -\infty < \gamma < t,$$

with

$$R_{YY}(t + \epsilon, \gamma) = E[Y(t + \epsilon) Y(\gamma)] \quad \text{and} \quad R_{YY}(\lambda, \gamma) = E[Y(\lambda) Y(\gamma)].$$

This reduces to

$$R_{YY}(t + \epsilon, \gamma) = \int_{-\infty}^{t} a(\lambda) R_{YY}(\lambda, \gamma) \, d\lambda, \quad -\infty < \gamma < t. \quad (11.20)$$

The resulting mean-square error is found from the inner product of the error and the quantity to be estimated. That is,

$$\overline{e^2} = E\left\{ \left[Y(t + \epsilon) - \int_{-\infty}^{t} a(\lambda) Y(\lambda) \, d\lambda \right] Y(t + \epsilon) \right\}$$

$$= R_{YY}(t + \epsilon, t + \epsilon) - \int_{-\infty}^{t} a(\lambda) R_{YY}(\lambda, t + \epsilon) \, d\lambda. \quad (11.21)$$

The optimum weighting function $a(\lambda)$ is the one that satisfies Eq. (11.20). That is, we must solve this integral equation in order to find the optimum

$a(\lambda)$. Having found this $a(\lambda)$, we may then solve Eq. (11.21) for the minimum mean-square error.

Notes. a) Equations (11.20) and (11.21) are true for general stochastic processes X and Y. When both processes are stationary, the autocorrelation functions depend only on time differences. Thus Eqs. (11.20) and (11.21) become

$$R_{YY}(t + \epsilon - \gamma) = \int_{-\infty}^{+\infty} a(\lambda) R_{YY}(\lambda - \gamma) \, d\lambda$$

and

$$\overline{e^2} = R_{YY}(0) - \int_{-\infty}^{t} a(\lambda) R_{YY}(\lambda - t - \epsilon) \, d\lambda.$$

b) Double subscripts are used on the autocorrelation functions, as opposed to single subscripts used on the autocorrelation functions in the previous chapter. This is to distinguish between autocorrelation which involves one process and cross correlation which involves two processes. Thus the cross correlation between processes X and Y is written as

$$R_{XY}(t_1, t_2) = E[X(t_1) Y(t_2)].$$

Solution of the Filtering Problem

As above, we find the solution to the filtering problem by choosing those weights $a(\lambda)$ that make the error orthogonal to the data. We find the solution as follows:

$$E\left\{\left[X(t) - \int_{-\infty}^{t} a(\lambda) Y(\lambda) \, d\lambda\right] Y(\gamma)\right\} = 0, \qquad -\infty < \gamma < t,$$

or

$$R_{XY}(t, \gamma) = \int_{-\infty}^{t} a(\lambda) R_{YY}(\lambda, \gamma) \, d\lambda, \qquad -\infty < \gamma < t, \qquad (11.22)$$

and the resulting mean-square error is

$$\overline{e^2} = E\left\{\left[X(t) - \int_{-\infty}^{t} a(\lambda) Y(\lambda) \, d\lambda\right] X(t)\right\}$$

$$= R_{XX}(t, t) - \int_{-\infty}^{t} a(\lambda) R_{YX}(\lambda, t) \, d\lambda. \qquad (11.23)$$

Notes. a) Again we note that the optimum weights $a(\lambda)$ are those that satisfy Eq. (11.22).

b) In the case of stationarity, the equations reduce to

$$R_{XY}(t - \gamma) = \int_{-\infty}^{t} a(\lambda) R_{YY}(\lambda - \gamma) \, d\lambda$$

and

$$\overline{e^2} = R_{XX}(0) - \int_{-\infty}^{t} a(\lambda) R_{YX}(\lambda - t) \, d\lambda.$$

Linear mean-square estimation theory has been used to arrive at a solution for the optimum weights $a(\lambda)$. Unfortunately, the solution is not complete. Suppose the correlation functions R_{XY} and R_{YY} in Eq. (11.22) are given. Then we must solve this integral equation for $a(\lambda)$. The solution of Eqs. (11.22) and (11.20) is the subject of the next chapter.

PROBLEMS

1. Two random variables X and Y are defined on the sample space given in Fig. 11.5. There are three possible experimental outcomes, ζ_1, ζ_2, and ζ_3, and they occur with equal probability.

x	-2	0	1
y	1	0	1
ζ	ζ_1	ζ_2	ζ_3

Fig. 11.5

a) Find the mean-square estimate of X.
b) Find the mean-square estimate of X given $Y = 1$.
c) Find the mean-square estimate of X given $Y = 0$.
d) Find the linear function $X = aY + b$ that will give the best linear mean-square estimate of X given Y.

2. Repeat Problem 1 for the probabilities

$$P(\zeta_1) = P(\zeta_2) = \tfrac{1}{4}, \qquad P(\zeta_3) = \tfrac{1}{2}.$$

3. The random variables X and N are defined as in Example 5. X is a random variable that assumes any of the values $-2, -1, 0, 1, 2$ with equal probability. Let $Y = X + N$, where N is another random variable, statistically independent from X, that assumes any of the values $-1, 0, 1$ with equal probability.

a) Calculate the mean-square estimate of X.
b) Calculate the mean-square estimate of X given $Y = -3$.
c) Calculate the mean-square estimate of X given $Y = -2$, then $-1, 0, 1, 2, 3$.
d) Calculate the average square error.

4. Find the linear mean-square estimate in Problem 3.

5. Let N be a gaussian random variable with zero mean and unit variance. Repeat all parts of Problem 3 except (d).

6. Given three observations, Y_1, Y_2, Y_3 in Problem 3, find the linear mean-square estimate.

7. Find the mean-square error in Problem 3 for the nonlinear estimate, and compare your answer to the mean-square error in Problem 4.

8. Generalize Problem 4 for random variables with nonzero mean. Let X take on the values 1, 2, 3, 4, 5 with equal probability (instead of $-2, -1, 0, 1, 2$), and repeat Problem 4.

FURTHER READING

1. A. Papoulis, *Probability, Random Variables, and Stochastic Processes*, McGraw-Hill, New York, 1965.
2. W. A. Porter, *Modern Foundations of Systems Engineering*, Macmillan, New York, 1966.

The approach to linear mean-square estimation taken here is based on the concept of a linear vector space. This is also the approach taken by Papoulis. Porter provides a good introduction to this subject.

12
OPTIMUM SYSTEMS

12.1 INTRODUCTION

In Chapter 11 we introduced the concept of *linear* mean-square estimation and applied it to several situations. We derived some simple relationships that related the optimum weights to the data and allowed us to calculate the mean-square error with ease. See, for example, Eqs. (11.15) and (11.16). We then applied this same concept to continuous data to solve the prediction problem and the filtering problem. In both cases we can view the optimum weights $a(t)$ as the impulse response of a filter, say $h(t)$.

From this viewpoint, then, the input data $y(t)$ is an electrical signal, the optimum weights $h(t)$ represent the filter impulse response, and the estimate $x(t)$ is the output of this filter. This viewpoint places the restriction of physical realizability (causalty) on $h(t)$; that is, $h(t) = 0$ for $t < 0$.

Let us quickly point out that other viewpoints are possible. For example, a digital computer may be used to operate on the data, in which case $h(t)$ need not be causal. (In the computer program, t is simply a parameter, and need not always progress in an increasing direction, as it does in an electronic filter.) Wainstein and Zubakov [3] call this the type I filter. In this filter the data may be stored as in a computer, and then operated on in an optimum fashion. The type II filter is the causal electronic filter which has data available from the infinite past, and the type III filter is the causal filter which has data available from the finite past (say from $t - T$ to t). We will use this classification system in the following discussion.

In the pure prediction problem, the input is assumed to be unperturbed signal (no noise), and the quantity to be estimated is the value of the signal at some future time. In the filtering problem, the input is the sum of signal plus noise, and the quantity to be estimated is the present value of signal. Since identical procedures are used in the solution of both problems, it is convenient to combine them. Thus suppose that the input signal is the sum of signal plus noise,

$$y(t) = x(t) + n(t),$$

and the quantity to be estimated is $x(t)$. The $x(t)$ may be the signal itself, $x(t) = s(t)$, which gives us the smoothing (filtering) problem; or $x(t)$ may be the signal at some future time, $x(t) = s(t + \epsilon)$ (prediction); or in general $x(t)$ may be any function of the signal (such as the derivative, integral, etc.).

We now investigate several related, progressively more complex problems.

12.2 THE OPTIMUM TYPE I FILTER

First, assume that the data is available for all time, $-\infty < t < +\infty$, and that a type I system with infinite storage capacity is used. Then, from Chapter 11, our problem is to find the $h(t)$ that makes the error orthogonal to the data. This will be the optimum $h(t)$.

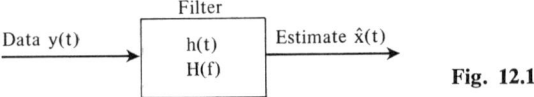

Fig. 12.1

Refer to Fig. 12.1. The data is $y(t)$ and the estimate is the output of the type I filter,

$$\hat{x}(t) = \int_{-\infty}^{+\infty} h(\lambda) y(t - \lambda) \, d\lambda,$$

so the error is

$$e(t) = x(t) - \int_{-\infty}^{+\infty} h(\lambda) y(t - \lambda) \, d\lambda.$$

Setting the error orthogonal to the data yields

$$E\left\{\left[x(t) - \int_{-\infty}^{+\infty} h(\lambda) y(t - \lambda) \, d\lambda\right] y(\gamma)\right\} = 0, \qquad -\infty < \gamma < +\infty,$$

or

$$R_{XY}(t, \gamma) = \int_{-\infty}^{+\infty} h(\lambda) R_{YY}(t - \lambda, \gamma) \, d\lambda, \qquad -\infty < \gamma < +\infty. \quad (12.1)$$

Now we are at the point where we stopped in Chapter 11. The optimum $h(t)$ is the one that satisfies Eq. (12.1). Fortunately, for the type I system the solution is simple if the input process $y(t)$ is stationary. Making this assumption allows us to write Eq. (12.1) as

$$R_{XY}(t - \gamma) = \int_{-\infty}^{+\infty} h(\lambda) R_{YY}(t - \lambda - \gamma) \, d\lambda, \qquad -\infty < \gamma < +\infty.$$

Let $\tau = t - \gamma$. Then

$$R_{XY}(\tau) = \int_{-\infty}^{+\infty} h(\lambda) R_{YY}(\tau - \lambda) \, d\lambda, \qquad -\infty < \tau < +\infty.$$

12.2 THE OPTIMUM TYPE I FILTER

We recognize the integral in this equation as the convolution of h with R_{YY}. Taking transforms of both sides yields

$$G_{XY}(f) = H(f)G_{YY}(f).$$

[$G_{XY}(f)$ is defined in the note below]. The transfer function of the optimum filter in the frequency domain is therefore

$$H(f) = \frac{G_{XY}(f)}{G_{YY}(f)}. \tag{12.2}$$

This is the general solution for the type I filter with stationary input. The most common situation is for the input to be the sum of statistically independent signal and noise,

$$y(t) = s(t) + n(t),$$

and the desired quantity is signal,

$$x(t) = s(t).$$

If either signal or noise has zero mean, then the cross terms $G_{SN}(f)$ and $G_{NS}(f)$ are zero, so that

$$H(f) = \frac{G_{SS}(f)}{G_{SS}(f) + G_{NN}(f)}. \tag{12.3}$$

Notes. a) Power spectral densities were introduced in Chapter 10 and defined by Eq. (10.5) as

$$G_{XX}(f) = E\left[\lim_{T \to \infty} \frac{1}{2T} |X_T(f, \zeta)|^2\right].$$

Now we must extend this definition to cross spectral densities such as $G_{XY}(f)$ in Eq. (12.2). Since

$$|X_T(f)|^2 = X_T(f)X_T(-f),$$

we define the cross spectral density as

$$G_{XY}(f) = E\left[\lim_{T \to \infty} \frac{1}{2T} X_T(f, \zeta)Y_T(-f, \zeta)\right]. \tag{12.4}$$

b) Several assumptions were made in deriving Eq. (12.3), and these should be noted. They are: (1) the system need not be causal and must have infinite storage (type I); (2) the data $y(t)$ and the desired information $x(t)$ both are processes with zero mean; (3) the input is wide-sense stationary, so the correlation functions depend on the differences in their arguments; (4) $y(t)$ is the sum of statistically independent signal plus noise, and one of them must have zero mean. [In view of (2), this implies that *both* must have zero mean.]

Mean-Square Error for Type I Systems

The inner product of $e(t)$ and $x(t)$ gives the mean-square error as

$$\overline{e^2} = E\left\{\left[x(t) - \int_{-\infty}^{+\infty} h(\lambda)y(t-\lambda)\,d\lambda\right]x(t)\right\}.$$

Making the assumptions noted above yields

$$\overline{e^2} = R_{XX}(0) - \int_{-\infty}^{+\infty} h(\lambda)R_{YX}(-\lambda)\,d\lambda. \tag{12.5}$$

This is one expression for the mean-square error in terms of correlation functions. We can derive one in terms of power spectra by setting

$$z(\tau) = R_{XX}(\tau) - \int_{-\infty}^{+\infty} h(\lambda)R_{YX}(\tau-\lambda)\,d\lambda, \tag{12.6}$$

so that

$$\overline{e^2} = z(0).$$

Taking transforms of both sides of Eq. (12.6) yields

$$Z(f) = G_{XX}(f) - H(f)G_{YX}(f).$$

Now with $z(\tau)$ given by

$$z(\tau) = \int_{-\infty}^{+\infty} Z(f)e^{j2\pi f\tau}\,df,$$

the mean-square error $z(0)$ is

$$\overline{e^2} = z(0) = \int_{-\infty}^{+\infty} [G_{XX}(f) - H(f)G_{YX}(f)]\,df. \tag{12.7}$$

Note. Equations (12.5) and (12.7) are valid for assumptions (1), (2), and (3) in the preceding note. For assumption (4), the error is given in terms of G_{SS} and G_{NN} as

$$\overline{e^2} = \int_{-\infty}^{+\infty} \frac{G_{SS}(f)G_{NN}(f)}{G_{SS}(f) + G_{NN}(f)}\,df. \tag{12.8}$$

We derived this equation by substituting Eq. (12.2) into Eq. (12.7) and combining terms. If you wish to attempt this algebra, you must recognize two facts: first $G_{XY}(f) = G_{YX}(-f)$ and, second, $G_{XX}(f) = G_{XX}(-f)$.

Examples

1. For the assumptions made in the preceding note, let the signal and noise power spectral densities be given as in Fig. 12.2. Then Eq. (12.3) gives $H(f) = 1$ where signal is present, and zero elsewhere. The mean-square error is zero because

$$H(f)G_{SS}(f) = G_{SS}(f),$$

12.2 THE OPTIMUM TYPE I FILTER 243

Fig. 12.2

so Eq. (12.7) gives

$$\overline{e^2} = \int_{-\infty}^{+\infty} [G_{SS}(f) - G_{SS}(f)] \, df = 0.$$

2. Consider the common AM radio transmitter. Bendat [5] points out that the exponential-cosine autocorrelation is a reasonable model for many signals, including modulated sine waves of random amplitude and phase. Thus suppose that the input to our receiver is signal plus white noise, where the signal autocorrelation is

$$R_{SS}(\tau) = Ae^{-a|\tau|}\cos \omega_0 \tau$$

and the noise autocorrelation is

$$R_{NN}(\tau) = \eta\delta(\tau).$$

Taking transforms, the associated power spectral densities are

$$G_{SS}(f) = \frac{Aa}{a^2 + (\omega - \omega_0)^2} + \frac{Aa}{a^2 + (\omega + \omega_0)^2},$$

$$G_{NN}(f) = \eta.$$

Note. The transform of $R_{SS}(\tau)$ is easily found by first noting that

$$e^{-a|\tau|} \leftrightarrow \frac{2a}{a^2 + \omega^2},$$

and then applying the modulation property, Eq. (3.16).

The functions G_{SS} and G_{NN} are plotted in Fig. 12.3. The actual signal spectrum out of an AM transmitter may be somewhat different from this, but this is a reasonable model to use. Let us now find the optimum receiver for such an input.

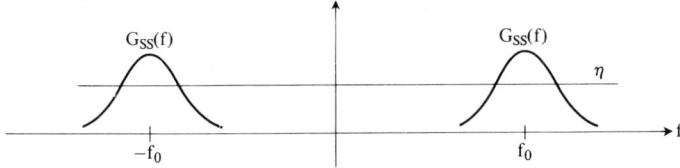

Fig. 12.3

For the situation shown in Fig. 12.3 ($f_0 \gg a$), the hump in G_{SS} at $f = -f_0$ contributes essentially nothing to the hump at $f = +f_0$. For this situation then, the optimum transfer function $H(f)$ is given by (see Eq. 12.3)

$$H(f) = \begin{cases} \dfrac{\dfrac{Aa}{a^2 + (\omega - \omega_0)^2}}{\eta + \dfrac{Aa}{a^2 + (\omega - \omega_0)^2}} = \dfrac{Aa}{Aa + \eta a^2 + \eta(\omega - \omega_0)^2}, & f > 0, \\[2em] \dfrac{\dfrac{Aa}{a^2 + (\omega + \omega_0)^2}}{\eta + \dfrac{Aa}{a^2 + (\omega + \omega_0)^2}} = \dfrac{Aa}{Aa + \eta a^2 + \eta(\omega + \omega_0)^2}, & f < 0, \end{cases} \quad (12.9)$$

which is two humps similar in shape to those for G_{SS} shown in Fig. 12.3, and in close agreement to the actual IF bandpass filters in a common AM receiver.

The mean-square error is not zero, since noise is present at all frequencies at which signal is present. The mean-square error is, from Eq. (12.8),

$$\overline{e^2} = 2 \int_0^\infty \frac{Aa\eta}{Aa + \eta a^2 + \eta(\omega - \omega_0)^2} \frac{d\omega}{2\pi}.$$

Note. This filter $H(f)$ may or may not be physically realizable. The point is that in deriving Eq. (12.3) we have taken no precautions to ensure physical realizability.

Can you determine whether $H(f)$ given by Eq. (12.9) is physically realizable? That is, is $h(t)$ zero for $t < 0$?

12.3 THE OPTIMUM TYPE II FILTER

We now require causalty: the impulse response of the optimum filter must be zero for $t < 0$. The entire past history of the input is assumed to be available, i.e., $y(\lambda)$ is given for $-\infty < \lambda < t$.

Refer to Fig. 12.1. The estimate is now given by

$$\hat{x}(t) = \int_0^\infty h(\lambda) y(t - \lambda) \, d\lambda,$$

so the error is

$$e(t) = x(t) - \int_0^\infty h(\lambda) y(t - \lambda) \, d\lambda.$$

12.3 THE OPTIMUM TYPE II FILTER

Setting the error orthogonal to the data yields

$$E\left\{\left[x(t) - \int_0^\infty h(\lambda)y(t-\lambda)\,d\lambda\right]y(\gamma)\right\} = 0, \qquad -\infty < \gamma < t,$$

or

$$R_{XY}(t,\gamma) = \int_0^\infty h(\lambda)R_{YY}(t-\lambda,\gamma)\,d\lambda, \qquad -\infty < \gamma < t.$$

Now assume stationarity and set $\tau = t - \gamma$. Then

$$R_{XY}(\tau) = \int_0^\infty h(\lambda)R_{YY}(\tau-\lambda)\,d\lambda, \qquad 0 < \tau < \infty. \tag{12.10}$$

Notes. a) Equation (12.10) is referred to in the literature as the Wiener-Hopf integral equation of the first kind.

b) Since $h(\lambda) = 0$ for $\lambda < 0$, the above integral is equal to

$$\int_{-\infty}^{+\infty} h(\lambda)R_{YY}(\tau-\lambda)\,d\lambda.$$

The only difference between Eq. (12.10) and the type I filter equation is the restriction on τ, $0 < \tau < \infty$. Unfortunately, this is a big difference, for Eq. (12.10) is not a convolution integral with this restriction, and the solution becomes complicated.

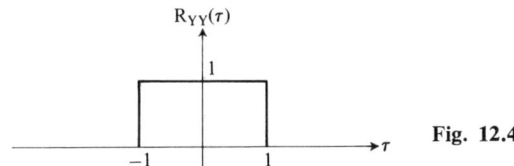

Fig. 12.4

This point can be illustrated simply by an example. Suppose $h(\tau) = u(\tau)$ (unit step) and $R_{YY}(\tau)$ is the square pulse shown in Fig. 12.4. Without the restriction $0 < \tau < \infty$, the integral in Eq. (12.10) would yield the function $R_{XY}(\tau)$ shown in Fig. 12.5(a). With the restriction $0 < \tau < \infty$, the function $R_{XY}(\tau)$ for $\tau < 0$ is zero, as shown in Fig. 12.5(b).

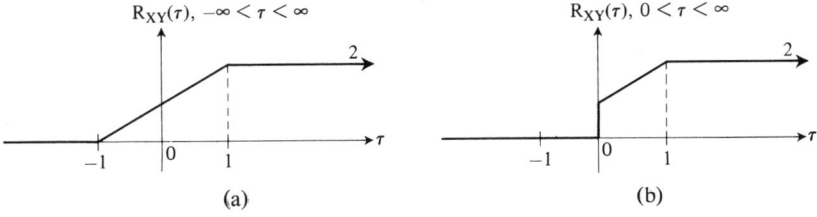

Fig. 12.5

Solution of Eq. (12.10)

The difficulty is encountered because Eq. (12.10) is valid only for $\tau > 0$. Why don't we write another equation that is identical to (12.10) for $\tau > 0$, but is valid for all τ? This artifice will then enable us to find a solution. Let us rewrite Eq. (12.10) as

$$\int_{-\infty}^{+\infty} h(\lambda) R_{YY}(\tau - \lambda)\, d\lambda - R_{XY}(\tau) = a(\tau), \qquad (12.11)$$

where $a(\tau) = 0$ for $\tau > 0$. Since the relationship between the two terms in Eq. (12.10) is unspecified for $\tau < 0$, we know only that $a(\tau)$ is not restricted to zero for $\tau < 0$. Taking the transform of Eq. (12.11), we have

$$H(s)G_{YY}(s) - G_{XY}(s) = A(s). \qquad (12.12)$$

Suppose $h(t) = 0$ for $t < 0$. Then we call $h(t)$ a positive time function. Also, if $a(t) = 0$ for $t > 0$, then $a(t)$ is a negative time function. Now any stable positive time function whose transform $H(s)$ can be written as a ratio of polynomials in s has all its poles in the left half of the s-plane (LHP). Likewise, if $A(s)$ is the transform of a stable negative time function, then $A(s)$ has all its poles in the right half of the s-plane (RHP).

Our aim is to split Eq. (12.12) into two parts, one of which is valid for $\tau > 0$ and therefore represents the solution of Eq. (12.10), and the other is valid for $\tau < 0$ and therefore is useless to us. To accomplish this, we factor G_{YY} into

$$G_{YY}(s) = G_{YY}^{+}(s) G_{YY}^{-}(s),$$

where $G_{YY}^{+}(s)$ corresponds to a positive time function with all its poles and zeros in the LHP. Now we can rewrite Eq. (12.12) as

$$H(s) G_{YY}^{+}(s) G_{YY}^{-}(s) - G_{XY}(s) = A(s)$$

or

$$H(s) G_{YY}^{+}(s) - \frac{G_{XY}(s)}{G_{YY}^{-}(s)} = \frac{A(s)}{G_{YY}^{-}(s)}.$$

Notes. a) Since $G_{YY}^{-}(s)$ has all its zeros in the RHP, the term $A(s)/G_{YY}^{-}(s)$ has all its poles in the RHP and corresponds to a negative time function.

b) $H(s) G_{YY}^{+}(s)$ has all its poles in the LHP and corresponds to a positive time function.

Now consider $G_{XY}(s)/G_{YY}^{-}(s)$ as the *sum* of two functions, one a positive time function (poles in LHP) and the other a negative time function. Then

$$\frac{G_{XY}(s)}{G_{YY}^{-}(s)} = \left[\frac{G_{XY}(s)}{G_{YY}^{-}(s)}\right]^{+} + \left[\frac{G_{XY}(s)}{G_{YY}^{-}(s)}\right]^{-},$$

where [ ]$^+$ denotes the positive time part of [ ]. We now arrive at our solution. The part of Eq. (12.12) that is valid for $\tau > 0$ is

$$H(s)G_{YY}^+(s) - \left[\frac{G_{XY}(s)}{G_{YY}^-(s)}\right]^+ = 0$$

or

$$H(s) = \frac{1}{G_{YY}^+(s)}\left[\frac{G_{XY}(s)}{G_{YY}^-(s)}\right]^+, \quad (12.13)$$

which is the optimum realizable type II filter.

Note. A similar derivation of Eq. (12.13) is presented in most of the references for this chapter. The derivation here is due to Brown and Nilsson [2].

Mean-Square Error for Type II Systems

The inner product of the error and the quantity we are estimating gives the mean-square error:

$$\overline{e^2} = E\left\{\left[x(t) - \int_0^\infty h(\lambda)y(t-\lambda)\,d\lambda\right]x(t)\right\}. \quad (12.14)$$

Expressions for the mean-square error differ according to the nature of $x(t)$, the quantity to be estimated. We will consider filtering and prediction cases.

Let us first consider filtering. In this case $x(t) = s(t)$, and the data is the sum of statistically independent stationary signal and noise:

$$y(t) = s(t) + n(t).$$

Equation (12.14) then becomes

$$\overline{e^2} = R_{SS}(0) - \int_0^\infty h(\lambda)R_{YS}(-\lambda)\,d\lambda. \quad (12.15)$$

Note the difference between this expression and the mean-square error for the type I system, Eq. (12.5). Here, the lower limit on the integral is zero instead of $-\infty$, as in Eq. (12.5). Since we are subtracting less from $R_{SS}(0)$ in the type II system, the mean-square error is larger.

Since $h(\lambda) = 0$ for $\lambda < 0$, the lower limit on the integral in Eq. (12.15) can be changed to $-\infty$ without affecting the value of the integral. Therefore the procedure used in arriving at Eq. (12.8) can be used here to obtain the mean-square error in terms of spectral densities. This is left as an exercise for the reader.

Now let us consider prediction. In this case $x(t) = s(t + \epsilon)$ and $y(t) = s(t)$. There is no noise. Equation (12.14) then becomes

$$\overline{e^2} = R_{SS}(0) - \int_0^\infty h(\lambda)R_{SS}(-\epsilon - \lambda)\,d\lambda. \quad (12.16)$$

We will study this special case in the next section.

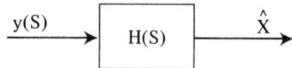

Fig. 12.6

Example 3. Consider the low-pass equivalent of Example 2. That is, suppose the input to our receiver is signal plus white noise and the output is the estimate $\hat{X}$ (see Fig. 12.6). The signal and noise autocorrelation functions are given by

$$R_{SS}(\tau) = Ae^{-a|\tau|} \quad \text{and} \quad R_{NN}(\tau) = \eta\delta(\tau).$$

Taking transforms, we find that the power spectral densities are

$$G_{SS}(\omega) = \frac{2aA}{a^2 + \omega^2} \quad \text{and} \quad G_{NN}(\omega) = \eta.$$

To keep from cluttering up the equations, let $\eta = a = 1$ and $2aA = 1$. The power spectral densities as a function of s are

$$G_{SS}(s) = \frac{-1}{-1 + s^2} \quad \text{and} \quad G_{NN}(s) = 1.$$

First, the optimum type I filter is, from Eq. (12.3),

$$H_1(s) = \frac{G_{SS}(s)}{G_{SS}(s) + G_{NN}(s)} = \frac{-1}{s^2 - 2}.$$

Using Eq. (12.8) to calculate the mean-square error for this type I system, we get

$$\overline{e^2} = \frac{1}{2\sqrt{2}} \approx 0.354.$$

But $H_1(s)$ is not physically realizable. One pole is in the RHP at $s = \sqrt{2}$.

The optimum type II system is found by use of Eq. (12.13). The steps are shown below:

$$G_{YY}(s) = G_{SS}(s) + G_{NN}(s) = \frac{-1}{-1 + s^2} + 1$$

$$= \frac{s^2 - 2}{s^2 - 1} = \frac{(s + \sqrt{2})(s - \sqrt{2})}{(s + 1)(s - 1)}.$$

The function G_{YY} is factored into two parts, so that

$$G_{YY}^+(s) = \frac{s + \sqrt{2}}{s + 1} \quad \text{and} \quad G_{YY}^-(s) = \frac{s - \sqrt{2}}{s - 1}.$$

The function $G_{XY}(s)$ in Eq. (12.13) is

$$G_{XY}(s) = G_{SS}(s) = \frac{-1}{-1 + s^2} = \frac{-1}{(s + 1)(s - 1)},$$

so that
$$\frac{G_{XY}(s)}{G_{YY}^-(s)} = \frac{-1}{(s+1)(s-\sqrt{2})} = \frac{1/(1+\sqrt{2})}{s+1} + \frac{-1/(1+\sqrt{2})}{s-\sqrt{2}}.$$

The function that corresponds to the positive time part of this function is the first term in the above partial fraction expansion; that is,
$$\left[\frac{G_{XY}(s)}{G_{YY}^-(s)}\right]^+ = \frac{1/(1+\sqrt{2})}{s+1}.$$

Putting this together with $1/G_{YY}^+(s)$ gives the optimum type II filter as
$$H_2(s) = \frac{1/(1+\sqrt{2})}{s+\sqrt{2}}.$$

Equation (12.15) can be used to calculate the mean-square error for this type II system. The impulse response is
$$h_2(t) = \frac{1}{1+\sqrt{2}} e^{-\sqrt{2}t}, \quad t > 0,$$

so that
$$\overline{e^2} = \frac{1}{2} - \int_0^\infty \frac{1}{1+\sqrt{2}} e^{-\sqrt{2}\lambda} \frac{1}{2} e^{-\lambda}\, d\lambda = \frac{1}{2} - \frac{1}{2(1+\sqrt{2})^2} \approx 0.415.$$

Note. This type II error of 0.415 is somewhat larger than the type I error of 0.354, as it must be.

12.4 PREDICTION IN THE ABSENCE OF NOISE

We have just presented the general theory of filtering and predicting in the presence of noise. A special case of this is prediction when there is no noise. That is, the input to our system is the past signal $s(\lambda)$, $-\infty < \lambda < t$, and the output is the estimate $\hat{s}(t+\epsilon)$.

We will now use a different approach to find the optimum system. This approach is due to Bode and Shannon.

To begin our discussion, let us focus our attention on Eq. (12.10). The solution for the pure prediction filter is found by setting the noise equal to zero and setting $x(t)$ equal to $s(t+\epsilon)$. But instead of solving Eq. (12.10) in this manner, we will take a different approach, and this approach is based on the observation that the optimum filter $h(t)$ is specified in Eq. (12.10) by the second-order statistics of $s(t)$. That is, the only information about $s(t)$ used in Eq. (12.10) is its autocorrelation function. Equivalently, in the frequency domain the optimum filter is related to the signal through the power spectral density function of $s(t)$.

Any other signal $g(t)$ with the same spectral density (and therefore the same autocorrelation) as $s(t)$ will use the same optimum filter $h(t)$ in prediction. This important but simple observation is the basis of the Bode-Shannon approach.

Let us assume that we have white noise $n(t)$ with zero mean present at point a in Fig. 12.7. The signal at point a is passed through filter $H_1(\omega)$ to obtain $g(t)$ at point b. Given that $g(t)$ has the same spectral density as $s(t)$, what is the optimum operation we must perform on it in order to estimate $g(t + \epsilon)$? Note that this same operation is the optimum thing we must do to $s(t)$ in order to estimate $s(t + \epsilon)$. As it turns out, the optimum thing we must do to $g(t)$ at point b is to convert it back to white noise $n(t)$ (point c) by the inverse filter $H_1^{-1}(\omega)$, and then operate on $n(t)$ in an optimum fashion to obtain $\hat{s}(t + \epsilon)$. Our reasoning is as follows: Knowledge of $g(t)$ at point b in Fig. 12.7 is equivalent to knowledge of $n(t)$ at point a together with knowledge of $H_1(\omega)$. Hence we have lost nothing (and gained nothing) in going from point b to point c in the figure. Therefore, any optimum operation on $n(t)$ will be equivalent to the optimum operation on $g(t)$ [or $s(t)$].

Notes. a) We plan to operate on $n(t)$ at point c in order to estimate $s(t + \epsilon)$, not $n(t + \epsilon)$. After all, $n(t)$ is white noise and its present value is uncorrelated with any future value. Therefore the past history of $n(t)$ is of no use in estimating $n(t + \epsilon)$.

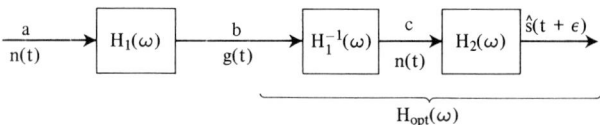

Fig. 12.7

b) The assumption that the spectral density of $s(t)$ can be obtained by filtering white noise restricts the type of signals we are discussing. They must be random (not predictable).

c) The inverse operation $H_1^{-1}(\omega)$ that converts $s(t)$ [or $g(t)$] to white noise may not be physically realizable. However, we only need to worry about the physical realizability of the combination $H_1^{-1}(\omega)H_2(\omega)$.

Now consider what we should do if we are given the noise up to the present time, $n(\lambda)$, $-\infty < \lambda < t$, and are asked to estimate $s(t)$, the present value of signal. The best we can do is to pass $n(t)$ through the filter $H_1(\omega)$ in Fig. 12.7 so that the output (point b) has the same spectral density as the signal. The output is not $s(t)$, but it is the best (mean-square) estimate of $s(t)$. With $h_1(t)$ the transform of $H_1(\omega)$, we have

$$\hat{s}(t) = \int_{-\infty}^{t} h_1(t - \lambda) n(\lambda) \, d\lambda.$$

12.4 PREDICTION IN THE ABSENCE OF NOISE

Similarly, the best estimate of $s(t + \epsilon)$ is

$$\hat{s}(t + \epsilon) = \int_{-\infty}^{t+\epsilon} h_1(t + \epsilon - \lambda)n(\lambda)\, d\lambda,$$

but the noise record is available only up to the present t and is unknown from t to $t + \epsilon$. That is, $n(\lambda)$ is known from $-\infty < \lambda < t$ and unknown from $t < \lambda < t + \epsilon$. If we break the integral into two parts, one over past values of λ and the other over future values, we have

$$\hat{s}(t + \epsilon) = \int_{-\infty}^{t} h_1(t + \epsilon - \lambda)n(\lambda)\, d\lambda + \int_{t}^{t+\epsilon} h_1(t + \epsilon - \lambda)n(\lambda)\, d\lambda. \quad (12.17)$$

We now maintain that the best mean-square estimate of $s(t + \epsilon)$ is obtained from the first integral alone. That is, if white noise $n(\lambda)$, $-\infty < \lambda < t$, is passed through a filter $h_1(t + \epsilon)$, $t > 0$, the output is the optimum estimate of $s(t + \epsilon)$. To see this, note that the mean value of the noise is zero. Since present (and past) values of noise are uncorrelated with future values, the best mean-square estimate for future values is zero. Hence, the best mean-square estimate of the second integral in Eq. (12.17) is zero.

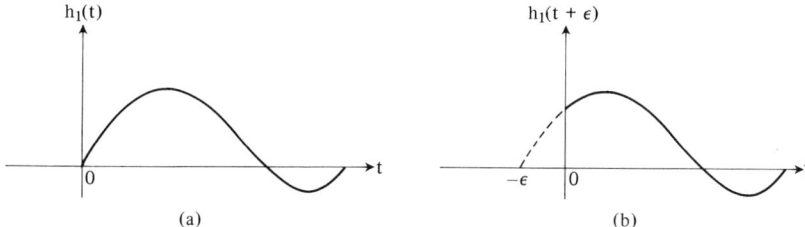

Fig. 12.8

Therefore the filter $h_2(t)$ in Fig. 12.7 is equal to $h_1(t + \epsilon)$, $t > 0$. The optimum filter to use on $s(t)$ is the cascaded pair $H_1^{-1}(\omega)H_2(\omega)$.

The filter $h_1(t)$ is the filter that converts white noise into a signal with the same spectrum as $s(t)$. If the impulse response $h_1(t)$ is as pictured in Fig. 12.8(a), then $h_2(t) = h_1(t + \epsilon)$, $t > 0$, is as shown in Fig. 12.8(b).

Note. $H_2(\omega)$ is not equal to $H_1(\omega)e^{-j\epsilon\omega}$, as one might suppose from the time-shifting property of Fourier transforms. For this property to hold, $h_1(t + \epsilon)$ cannot be restricted to zero for $t < 0$.

The following are the steps one must follow to construct the optimum predictor $h(t)$:

1. Determine $G_S(f)$, the spectrum of $s(t)$.
2. Set $G_S(f) = H_1(\omega)H_1^*(\omega)$ and solve for $H_1(\omega)$, the physically realizable filter that will convert white noise to a signal with the spectrum $G_S(f)$.

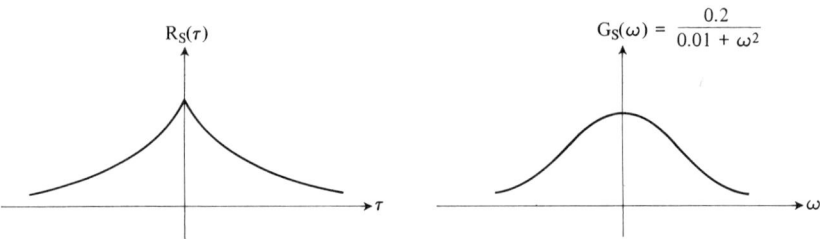

Fig. 12.9

3. Take the inverse Fourier transform of $H_1(\omega)$ to determine $h_1(t)$.
4. Set $h_2(t) = h_1(t + \epsilon)$, $t > 0$.
5. Take the transform of $h_2(t)$ to determine $H_2(\omega)$.
6. Determine the optimum filter $H(\omega)$ by

$$H(\omega) = H_1^{-1}(\omega) H_2(\omega) = \frac{H_2(\omega)}{H_1(\omega)}.$$

Example 4. Find the best realizable filter to estimate $s(t + 1)$ given that the signal has correlation $R_S(\tau) = e^{-0.1|\tau|}$. See Fig. 12.9.

Solution. The number of the steps below correspond to the steps given above for finding $h(t)$.

1. The power spectral density is the transform of $R_S(\tau)$ given by

$$G_S(\omega) = \frac{0.2}{0.01 + \omega^2}.$$

2. $G_S(\omega) = \dfrac{\sqrt{0.2}}{0.1 + j\omega} \dfrac{\sqrt{0.2}}{0.1 - j\omega} = H_1(\omega) H_1^*(\omega).$

Hence the physically realizable filter $H_1(\omega)$ is

$$H_1(\omega) = \frac{\sqrt{0.2}}{0.1 + j\omega}.$$

3. $h_1(t) = \displaystyle\int_{-\infty}^{+\infty} H_1(\omega) e^{j\omega t} \frac{d\omega}{2\pi} = \sqrt{0.2}\, e^{-0.1t}, \quad t > 0.$

4. $h_2(t) = \sqrt{0.2}\, e^{-0.1(t+1)}, \quad t > 0.$

5. The Fourier transform of $h_1(t + 1)$ is

$$H_2(\omega) = \int_0^\infty \sqrt{0.2}\, e^{-0.1} e^{-0.1t} e^{-j\omega t}\, dt = \frac{\sqrt{0.2}\, e^{-0.1}}{0.1 + j\omega}.$$

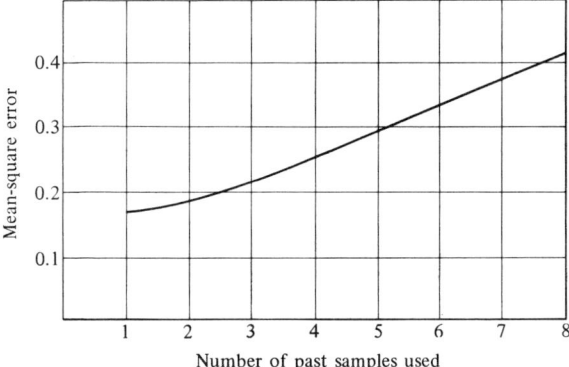

Fig. 12.10

6. The optimum realizable filter is

$$H(\omega) = \frac{H_2(\omega)}{H_1(\omega)} = e^{-0.1} \approx 0.9.$$

Hence the best mean-square estimate of $s(t+1)$ is the present value of $s(t)$ multiplied by 0.9. The past history of $s(t)$ is not used—only the present value. Thus we have

$$\hat{s}(t+1) = 0.9s(t).$$

Figure 12.10 shows the results of an experiment performed to illustrate this point. A set of correlated gaussian random numbers were generated, with correlation $R_S(\tau) = e^{-0.1|\tau|}$. The present and past values were used to estimate the next value, $s(t+1)$. When only the present value (times 0.9) was used, the mean-square error was 0.17. Any combination of past and present values only increased the error. The graph shows what happened to the error as more samples were used for one particular weighting scheme (all weights equal).

PROBLEMS

1. A stationary stochastic process $X(t)$ with zero mean has correlation

$$R_X(\tau) = \frac{\sin \pi \tau}{\pi \tau}.$$

One sample is taken at time t. It is $x(t) = 1.5$.
a) Find the linear estimate of $x(t + \frac{1}{2})$.
b) Find the linear estimate of $x(t + \frac{3}{2})$.
c) Calculate the mean-square error in (a).
d) Calculate the mean-square error in (b).

2. Now suppose you are given two values of x: $x(t) = 1.5$ and $x(t - \frac{1}{2}) = -0.5$. Repeat Problem 1.

3. Let $y(t) = s(t) + n(t)$, where the signal $s(t)$ and noise $n(t)$ are statistically independent with zero mean and autocorrelation functions

$$R_S(\tau) = e^{-0.1|\tau|} \quad \text{and} \quad R_N(\tau) = \delta(\tau).$$

Find the optimum type I (nonrealizable) filter used to estimate the present value of signal if $y(t)$ is available for $-\infty < t < +\infty$, and calculate the mean-square error.

4. Find the optimum type II (realizable) filter in Problem 3 used to estimate $s(t)$ at the present time if $y(t)$ is available from $-\infty$ to the present. Also find the mean-square error incurred by using this filter.

5. Find the optimum type II filter used to estimate signal one second in the future if the autocorrelation function is $R_S(\tau) = e^{-0.1|\tau|}$.

6. Let $y(t) = s(t) + n(t)$, where signal $s(t)$ and noise $n(t)$ are statistically independent with zero mean and autocorrelation functions

$$R_S(\tau) = e^{-0.1|\tau|} \quad \text{and} \quad R_N(\tau) = \frac{\sin \pi \tau}{\pi \tau}.$$

You are given just one measurement of $y(t)$. It is $y(0) = 1.5$. Estimate $s(0)$, and find the mean-square error.

7. In Problem 6, suppose you are given two measurements of $y(t)$. They are $y(0) = 1.5$ and $y(-1) = -0.5$. Estimate $s(0)$, and find the mean-square error.

8. Find the transfer function that will accomplish optimum linear prediction of a random signal $s(t)$ with spectral density

$$G_S(\omega) = \frac{4}{(\omega^2 + 1)(\omega^2 + 4)}.$$

FURTHER READING

1. A. PAPOULIS, *Probability, Random Variables, and Stochastic Processes*, McGraw-Hill, New York, 1965.
2. R. G. BROWN and J. W. NILSSON, *Linear Systems Analysis*, Wiley, New York, 1962.
3. L. A. WAINSTEIN and V. D. ZUBAKOV, *Extraction of Signals from Noise*, Prentice-Hall, Englewood Cliffs, N.J., 1962.
4. W. B. DAVENPORT, JR., and W. L. ROOT, *Random Signals and Noise*, McGraw-Hill, New York, 1958.
5. J. S. BENDAT, *Principles and Applications of Random Noise Theory*, Wiley, New York, 1958.

6. H. W. BODE and C. E. SHANNON, "A Simplified Derivation of Linear Least Square Smoothing and Prediction Theory," *Proc. IRE*, (April 1950), p. 417.

Each of References 1 through 5 provides a good discussion of all the material in this chapter. Section 12.4 is based on Reference 6.

The area under the standard normal curve from 0 to z (the shaded area) is $A(z)$.

Table of Normal Curve Areas

Examples. If Z is the standard normal random variable and $z = 1.54$, then

$$A(z) = P\{0 < Z < z\} = .4382,$$
$$P\{Z > z\} = .0618,$$
$$P\{Z < z\} = .9382,$$
$$P\{|Z| < z\} = .8764.$$

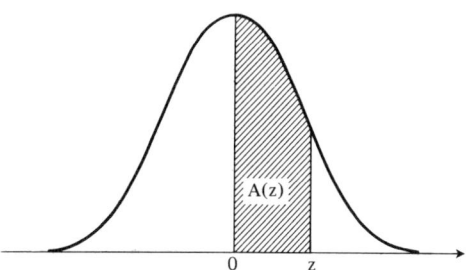

z	.00	.01	.02	.03	.04	.05	.06	.07	.08	.09
0.0	.0000	.0040	.0080	.0120	.0160	.0199	.0239	.0279	.0319	.0359
0.1	.0398	.0438	.0478	.0517	.0557	.0596	.0636	.0675	.0714	.0753
0.2	.0793	.0832	.0871	.0910	.0948	.0987	.1026	.1064	.1103	.1141
0.3	.1179	.1217	.1255	.1293	.1331	.1368	.1406	.1443	.1480	.1517
0.4	.1554	.1591	.1628	.1664	.1700	.1736	.1772	.1808	.1844	.1879
0.5	.1915	.1950	.1985	.2019	.2054	.2088	.2123	.2157	.2190	.2224
0.6	.2257	.2291	.2324	.2357	.2389	.2422	.2454	.2486	.2517	.2549
0.7	.2580	.2611	.2642	.2673	.2704	.2734	.2764	.2794	.2823	.2852
0.8	.2881	.2910	.2939	.2967	.2995	.3023	.3051	.3078	.3106	.3133
0.9	.3159	.3186	.3212	.3238	.3264	.3289	.3315	.3340	.3365	.3389
1.0	.3413	.3438	.3461	.3485	.3508	.3531	.3554	.3577	.3599	.3621
1.1	.3643	.3665	.3686	.3708	.3729	.3749	.3770	.3790	.3810	.3830
1.2	.3849	.3869	.3888	.3907	.3925	.3944	.3962	.3980	.3997	.4015
1.3	.4032	.4049	.4066	.4082	.4099	.4115	.4131	.4147	.4162	.4177
1.4	.4192	.4207	.4222	.4236	.4251	.4265	.4279	.4292	.4306	.4319
1.5	.4332	.4345	.4357	.4370	.4382	.4394	.4406	.4418	.4429	.4441
1.6	.4452	.4463	.4474	.4484	.4495	.4505	.4515	.4525	.4535	.4545
1.7	.4554	.4564	.4573	.4582	.4591	.4599	.4608	.4616	.4625	.4633
1.8	.4641	.4649	.4656	.4664	.4671	.4678	.4686	.4693	.4699	.4706
1.9	.4713	.4719	.4726	.4732	.4738	.4744	.4750	.4756	.4761	.4767
2.0	.4772	.4778	.4783	.4788	.4793	.4798	.4803	.4808	.4812	.4817
2.1	.4821	.4826	.4830	.4834	.4838	.4842	.4846	.4850	.4854	.4857
2.2	.4861	.4864	.4868	.4871	.4875	.4878	.4881	.4884	.4887	.4890
2.3	.4893	.4896	.4898	.4901	.4904	.4906	.4909	.4911	.4913	.4916
2.4	.4918	.4920	.4922	.4925	.4927	.4929	.4931	.4932	.4934	.4936
2.5	.4938	.4940	.4941	.4943	.4945	.4946	.4948	.4949	.4951	.4952
2.6	.4953	.4955	.4956	.4957	.4959	.4960	.4961	.4962	.4963	.4964
2.7	.4965	.4966	.4967	.4968	.4969	.4970	.4971	.4972	.4973	.4974
2.8	.4974	.4975	.4976	.4977	.4977	.4978	.4979	.4979	.4980	.4981
2.9	.4981	.4982	.4982	.4983	.4984	.4984	.4985	.4985	.4986	.4986
3.0	.4987	.4987	.4987	.4988	.4988	.4989	.4989	.4989	.4990	.4990

ANSWERS TO SELECTED PROBLEMS

ANSWERS TO SELECTED PROBLEMS

Chapter 1

2. a) $P = \frac{1}{2}$, $E = \infty$ b) $P = 0$, $E = 50$
   c) $P = 0$, $E = 1$ d) $P = \frac{1}{2}$, $E = \infty$

Chapter 2

3. a) $v(0) = 5$
   b) $v(1) = e^{-j2\omega_1} + e^{-j\omega_1} + 1 + e^{j\omega_1} + e^{j2\omega_1}$
   c) $v(t) = 2\cos 2\omega_1 t + 2\cos \omega_1 t + 1$
   d) $T = 1/f_1$ but f_1 is unknown.
4. $v(t) = 1 + \cos(\omega_1 t + \theta_1) + \frac{1}{2}\cos(2\omega_1 t + \theta_2)$
5. $g(t) = \frac{1}{2} + (2/\pi)\cos(\omega_1 t + \pi/2) + (1/\pi)\cos(2\omega_1 t + \pi)$

7. $Y_{\text{rms}} = 0.69$ V   8. $F_K = Ea\left(\dfrac{\sin K\pi a}{K\pi a}\right)$

9. a) $E = 2.23$ V, $a = 0.47$ b) $V_{\text{rms}} = 1.53$ V
10. b) $P_0 = 0.475$ W   c) $P_0 = 4.75(10)^{-4}$ W
11. $Y_{\text{rms}} = 0.9$ V
12. a) $C = 630$ pF   b) $V_{\text{rms}} = 0.86$ V
13. For $e_1(t)$, the output is zero. For $e_2(t)$, the output is $e_2(t)$.
14. $V_{\text{rms}} = 0.27$ V
16. a) $\frac{1}{2}$ b) $\frac{1}{2}$ c) $\frac{1}{4}$ d) $\frac{3}{4}$ e) $\frac{1}{2}$ f) $\frac{3}{8}$
19. $y(t) = 2\cos(2\pi 50 t + 45°)$
20. $\frac{1}{2}$ 21. b) $y(t) = (V_1/\pi)\cos(2\pi f_A t) + V_0/\pi$
22. $Y_{\text{rms}} = 0.012$ V

Chapter 3

1. a) $F(f) = 1/(a + j\omega)$ b) $F(f) = 1/(a - j\omega)$
   c) $F(f) = 2a/(a^2 + \omega^2)$ d) $F(f) = -2j\omega/(a^2 + \omega^2)$

ANSWERS TO SELECTED PROBLEMS

2. a) $V_1(\omega) = \dfrac{1}{j\omega} + \dfrac{e^{-j\omega} - 1}{-\omega^2}$ b) $V_2(\omega) = \left[\dfrac{1}{j\omega} + \dfrac{e^{-j\omega} - 1}{-\omega^2}\right][1 + e^{-j2\omega}]$

5. $s(t) = 2a \dfrac{\sin(2\pi t - 2)a}{(2\pi t - 2)a}$ 9. $V_1(\omega) = \dfrac{2\pi}{\Delta}\left(\dfrac{\cos\omega\Delta/2}{\omega^2 - \pi^2/\Delta^2}\right)$

10. a) $V(\omega) = \dfrac{(\sin\omega/4)^2}{(\omega/4)^2}$

c) content = 1, variation = 4, wiggliness = 16

14. $|G(f)|^2 = \dfrac{2E^2}{T^2\omega^4} - \dfrac{2E^2}{T^2\omega^4}\cos\omega T - \dfrac{2E^2}{T\omega^3}\sin\omega T + \dfrac{E^2}{\omega^2}$

16. a) 0.2 J   b) 0.95 J   c) 1 J

17. $|V(\omega)|^2 = \dfrac{4}{4 + \omega^2}$ 18. 0.33 J

Chapter 4

1. a) $v(t) = Ri(t) + L\dfrac{di}{dt}$ b) $H(s) = \dfrac{1/L}{S + R/L}$

c) $h(t) = \dfrac{1}{L}e^{(-R/L)t}, t > 0$

2. a) $\dfrac{dv}{dt} = R\dfrac{di}{dt} + \dfrac{i(t)}{C}$ b) $H(s) = \dfrac{s/R}{s + 1/RC}$

c) $h(t) = \dfrac{1}{R}\delta(t) - \dfrac{1}{R^2C}e^{-t/RC}, t > 0$

3. a) $H(s) = \dfrac{s(s + 1)}{2s^2 + 2s + 1}$ b) $h(t) = \tfrac{1}{2}\delta(t) - \dfrac{e^{-0.5t}}{2}\sin 0.5t, t > 0$

5. a) $\sigma > -R/L$ b) $\sigma > -1/RC$ c) $\sigma > -0.5$

8. $y(t) = \begin{cases} 0, & t < 0, \\ 2[1 - e^{-t}], & 0 < t < 1, \\ 2[e^{-(t-1)} - e^{-t}], & t > 1 \end{cases}$

10. $y(t) = \begin{cases} 0, & t < 0, \\ 2 - t - 2e^{-t}, & 0 < t < 1, \\ e^{-(t-1)} - 2e^{-t}, & t > 1 \end{cases}$

11. $v_2(t) = 5e^{-10t}, t > 0$

12. a) $v_2(t) = \tfrac{1}{2}[1 - e^{-10t}], t > 0$ b) $v_{2\text{Problem }12}(t) = \displaystyle\int_{-\infty}^{t} v_{2\text{Problem }11}(\lambda)\, d\lambda$

13. a) $v_2(t) = \tfrac{1}{2}t - \tfrac{1}{20} + \tfrac{1}{20}e^{-10t}, t > 0$

14. $v_2(t) = \tfrac{1}{4}t^2 - \tfrac{1}{20}t + \tfrac{1}{200} - \tfrac{1}{200}e^{-10t}, t > 0$

ANSWERS TO SELECTED PROBLEMS   263

16. a) $e^{-t}\sin t$ b) $10 + 3e^{-2t}$ c) $3e^{-2t} + e^{-t}\sin t$ d) 10
f) $s^2 + 2s + 2 = 0$

17. $u(t) - u(t - \tau)$ 18. $9v_1(t) = 8\dfrac{d^2v_2}{dt^2} + 88\dfrac{dv_2}{dt} + 80v_2(t)$

19. $\tfrac{1}{2}[1 - e^{-2t}]u(t)$ 20. $2v_1(t) = \dfrac{d^2v_2}{dt^2} + 4\dfrac{dv_2}{dt}$

Chapter 5

2. a) $A - B = \{-5, -4, -3, -2, -1, 0\}$
   b) $B - A = \emptyset$ c) $A \cup B = A$ d) $A \cap B = B$

4. a) 0.4   b) 1   c) 0.6   d) 0.6   e) 0   f) 0.3
   g) 0.5

5. a) $S = \{1, 2, 3, 4, 5, 6\}$ b) $\tfrac{4}{6}$ c) $\tfrac{2}{6}$ d) 0   e) $\tfrac{2}{6}$

6. a) $\tfrac{1}{4}$ b) $\tfrac{1}{2}$ c) $\tfrac{3}{4}$ d) 0

7. a) $S = \{x : 0 \leq x \leq 2\pi\}$ b) $\tfrac{1}{2}$ c) $3/2\pi$ d) $\pi/6$

11. a) $A \cup B = \{a, b, c, d, e, f, g, h, i\}$ b) $A \cap B = \{e, f, g\}$

12. a) S is the 4! different arrangements of the form elements $\{1, 2, 3, 4\}$.

13. $\tfrac{9}{64}$ 14. $\tfrac{175}{256}$

15. a) 0.5275   b) 0.4725   c) 0.491

16. $\tfrac{3}{5}$

17. b) $P(e|B) = \tfrac{3}{13}$ c) $P(e|R) = \tfrac{2}{5}$ d) 0.277

18. $\tfrac{1}{3}$ 19. a) $\tfrac{1}{6}$ b) $\tfrac{4}{9}$

20. $\tfrac{1}{15}$ 21. a) $\tfrac{1}{16}$ b) $\tfrac{5}{16}$

22. 0.4929   23. $\tfrac{3}{8}$

Chapter 6

2. a) $a = 1/b$ c) $\tfrac{1}{8}$

3. a) $F_X(b - \epsilon) - F_X(a)$ for small ϵ b) $F_X(b) - F_X(a - \epsilon)$ for small ϵ
   c) $F_X(b) - F_X(a)$

6. a) 0   b) $\tfrac{1}{2}$ c) $\tfrac{3}{4}$

7. a) $b = 2a$ b) $m_Y = 0$, $\sigma_Y^2 = 2/b^2$

8. a) $m_X = x_1 p_1 + x_2 p_2 + x_3 p_3$
   b) $\sigma_X = (x_1 - m_X)^2 p_1 + (x_2 - m_X)^2 p_2 + (x_3 - m_X)^2 p_3$

9. a) 2   b) $\tfrac{13}{3}$ c) $\tfrac{1}{4}$

10. a) 2   b) 8   c) 0.4013

11. a) $\frac{3}{8}$ b) $\frac{7}{8}$ 12. 0.5905
13. a) 0.2304   b) 0.9129   14. $-1/a$
18. a) 1   b) 0   c) $f_X(x)$ is uniform $(-1, 1)$ d) $\phi_Y(v) = e^{jav}\dfrac{\sin v}{v}$
19. b) 0.389

Chapter 7

2. $f_Y(y) = \dfrac{1}{a} f_X\left(\dfrac{y-b}{a}\right)$

3. $f_Y(y) = \dfrac{f_X\left(\sqrt{\dfrac{y-b}{a}}\right) + f_X\left(-\sqrt{\dfrac{y-b}{a}}\right)}{\left|2a\sqrt{\dfrac{y-b}{a}}\right|}, y > b$

5. a) $\frac{9}{20}$ b) $\frac{200}{3}$

6. a) $F_Y(y) = \begin{cases} 1, & y \geq 1, \\ \frac{1}{2}y, & 0 \leq y < 1, \\ 0, & y < 0 \end{cases}$ b) $F_Y(y) = \begin{cases} 0, & y < 0, \\ \frac{1}{2}\sqrt{y}, & 0 \leq y < 4, \\ 1, & y \geq 4 \end{cases}$

c) $F_Y(y) = \begin{cases} 0, & y < 0, \\ \frac{1}{2}, & 0 \leq y < 1, \\ 1, & y \geq 1 \end{cases}$ d) $F_Y(y) = \begin{cases} 1, & y \geq 1, \\ y, & 0 \leq y < 1, \\ 1, & y < 0 \end{cases}$

7. a) $F_Y(y) = \begin{cases} 0, & y < -1, \\ F_X(y), & -1 \leq y < 1, \\ 1, & y \geq 1 \end{cases}$

10. $f_Y(y) = \begin{cases} 0, & y < 0, \\ \dfrac{2e^{-y^2/2}}{\sqrt{2\pi}}, & y \geq 0 \end{cases}$

11. a) $\frac{1}{4}$ b) $\frac{1}{4}(1 - e^{-1/8})$
12. $f_\Theta(\theta) = 1/2\pi, 0 < \theta < 2\pi,$
    $f_R(r) = re^{-r^2/2}, 0 < r$
13. a) $\frac{1}{2}$

Chapter 8

1. a) 4   b) $\frac{5}{9}$ 8. 250   9. 0.8245
10. a) 0.8245   b) 0.5628   c) 0.8245   d) $1 - e^{-1/2}$.

ANSWERS TO SELECTED PROBLEMS   265

Chapter 9

1. c) $\frac{29}{8}$ d) $\frac{21}{8}$ e) $\frac{63}{8}$
f) $f_{X_1X_2}(1,5;t_1,t_2) = \frac{1}{8}$,
$f_{X_1X_2}(2,4;t_1,t_2) = \frac{1}{4}$,
$f_{X_1X_2}(6,2;t_1,t_2) = \frac{3}{8}$,
$f_{X_1X_2}(3,1;t_1,t_2) = \frac{1}{4}$

2. a) $m_X(t) = \begin{cases} 0, & t < 0, \\ \frac{1}{2}, & t > 0 \end{cases}$ b) $\frac{1}{2}$ c) $\frac{1}{4}$

3. $E[X(2)] = \frac{10}{3}$, $E[X(6)] = \frac{11}{3}$, $R_{XX}(2,6) = \frac{31}{3}$

5. a) $m_X(t) = \frac{1}{3}[1 + \sin t + \cos t]$
b) $R_X(t_1, t_2) = \frac{1}{3}[1 + \sin t_1 \sin t_2 + \cos t_1 \cos t_2]$

6. a) $e^{-|\tau|} + \cos 2\pi\tau$
b) $e^{-|\tau|} + \cos 2\pi\tau$
c) $e^{-|\tau|} - \cos 2\pi\tau$

11. a) 2   b) 1   c) 1   d) No   e) No

Chapter 10

1. $G_X(\omega) = \frac{T}{2} \frac{\sin^2 \omega T/4}{(\omega T/4)^2}$

4. a) 0   b) $t_1 t_2$

5. a) $m_X(t) = \frac{\sin \omega_2 t - \sin \omega_1 t}{\omega_2 t - \omega_1 t}$

b) $R_X(t_1, t_2) = \frac{1}{2(\omega_2 - \omega_1)} \left[\frac{\sin \omega_2 \tau - \sin \omega_1 \tau}{\tau} + \frac{\sin \omega_2 \lambda - \sin \omega_1 \lambda}{\lambda} \right]$,
where $\tau = t_2 - t_1$, $\lambda = t_2 + t_1$

6. a) 0

b) $R_Y(\tau) = \begin{cases} 1 - \frac{|\tau|}{T}, & |\tau| < T, \\ 0, & |\tau| > T \end{cases}$

7. a) $\frac{1}{6}$ b) $R_X(\tau) = \frac{1}{6}[2e^{-|\tau|/2} \cos(\sqrt{3}/2)\tau - e^{-|\tau|}]$

8. a) $G_Y(\omega) = \frac{1}{1+\omega^2}$
b) $R_Y(\tau) = \frac{1}{2}e^{-|\tau|}$

Chapter 11

1. a) $-\frac{1}{3}$ b) $-\frac{1}{2}$ c) 0   d) $\hat{X} = -\frac{1}{2}Y$
2. a) 0   b) 0   c) 0

3. a) 0   b) -2
c) $E(X|Y = -2) = -1.5$,
   $E(X|Y = -1) = -1$,
   $E(X|Y = 0) = 0$,
   $E(X|Y = 1) = 1$,
   $E(X|Y = 2) = 1.5$,
   $E(X|Y = 3) = 2$
d) $\overline{e^2} = \frac{7}{15}$
4. $\hat{X} = \frac{3}{4}Y$
5. a) 0   b) -1.8
c) $E(X|Y = -2) = -1.48$,
   $E(X|Y = -1) = -0.88$,
   $E(X|Y = 0) = 0$,
   $E(X|Y = 1) = 0.88$,
   $E(X|Y = 2) = 1.48$,
   $E(X|Y = 3) = 1.8$,
8. $\hat{X} = \frac{3}{4}(Y - 3) + 3$

Chapter 12

1. a) 0.96   b) -0.32 c) 0.594   d) 0.955
2. a) 1.94   b) -0.555 c) 0.32   d) 0.923
3. a) $H_1(\omega) = \dfrac{0.2}{0.21 + \omega^2}$ b) $\overline{e^2} = 0.218$
4. a) $h_2(t) = 0.358 e^{-\sqrt{0.21}\,t}$, $t > 0$ b) $\overline{e^2} = 0.82$
5. $H(\omega) = e^{-0.5}$ 6. $\hat{s}(0) = 0.75$, $\overline{e^2} = 0.5$
7. $\hat{s}(0) = 0.416$, $\overline{e^2} = 0.372$

INDEX

INDEX

additive system, 63
adjoint of a matrix, 177
almost periodic function, 8
aperiodic signals, 7
autocorrelation, 199
autocorrelation function, 106

Bayes' law, 97
Bendat, J. S., 243
Bernoulli distribution, 120
binary distribution, 120
binomial coefficient, 122
binomial distribution, 121
binomial expansion, 122
Bode-Shannon technique, 249
bounds on the spectrum, 47
Brown, R. G., 247

cell, 94
central limit theorem, 125, 162
central moment, 118
characteristic equation, 64
characteristic function, 127
Chebyshev's Inequality, 157
class, 85
cofactor, 175
collection, 85
column vector, 173
complement, 87, 88
complementary solution, 64
composite function, 134
conditional probability, 95

conditional probability distribution, 149, 152
content, for aperiodic waveforms, 48
   for periodic waveforms, 47
continuous distribution, 114
continuous random variable, 112
convergence of $\mathcal{L}$-transform, 69
convolution, 62, 70, 214
correlation coefficient, 149
correlation function, 190
covariance, 149, 170
covariance function, 190
covariance matrix, 170
Cramér, H., 164
cross spectral density, 241
cumulative distribution function (cdf), 108

determinants, 175
difference (set), 87
differentiation property, 129
discrete distribution, 113
discrete random variable, 112
disjoint sets, 89
dominance condition, 68
dot product, 14

element, 85
empty set, 88
energy, 7
energy spectral density, 50
ensemble, 184

INDEX

equality of sets, 89, 90
equivalence of sets, 89
equivalent events, 136, 203
ergodic process, 191, 193
Euler's formula, 30
events, 90
exhaustive, 94
expected value, 117
experiment, 92

filtering, 247
finite sets, 89
Fourier series, 10
   convergence, 12
   domain, 11
   properties, 21
Fourier transform, 39, 127
   domain, 40
   properties, 44
function, 10
function of a random variable, 134
functional, 14
fuzzy set, 85

gaussian conditional density, 166
gaussian distribution, 125
gaussian marginal density, 166
gaussian process, 190
gaussian random variable, 164
geometric vectors, 230
group, 85

identity matrix, 178
impulse response, 76
independence, 98
independent random variables, 147
infinite set, 89
inner product, 14
inner product space, 230
intersection, 86
inverse of a matrix, 178

Jacobs, I. M., 164
joint cdf, 145
joint moments, 148
joint pdf, 146
jointly distributed random variables, 144

Kotel'nikov, V. A., 4

Laplace transform, 42
law of large numbers, 159
limiter, 142
linear mean-square estimation, 228
linear vector space, 230

marginal distribution, 147, 148
matrices, 173
mean-square estimation, 224
mean-square value, 31
measure, 92
member, 85
minor, 175
modulation property, 128
moments, 117
moments of stochastic processes, 185
mutually exclusive events, 98

negative time function, 246
Nilsson, J. W., 247
normal distribution, 125
normalizing factor, 167

one-to-one correspondence, 89
operator, 10
orthogonal expansion, 13
   functions, 13
output power spectral density, 31

Parseval's theorem, 25
particular solution, 64
partition, 94
Pascal's triangle, 122
periodic functions, 8
point, 85
Polyphemus, 89
positive time function, 246
power, 6
power spectral density, 26, 199, 201
prediction, 234, 247
probability, 90, 91
probability density function (pdf), 109
   measure, 92
properties, of gaussian random
     variables, 169
   of matrices, 179
pulse signals, 7

random signals, 9
random variable, 107
 function of a, 134
 two-dimensional, 144
random vector, 144
range, 10
Rayleigh, Lord, 49
Rayleigh's theorem, 49
rms value, 31
row vector, 173

sample space, 90
scaling property, 129
set, 85
signals, almost periodic, 8
 classification of, 5
 energy, 7
 periodic, 8
 power, 6
 random, 9
sinusoidal distribution, 116
 derived, 141
space, 85
specifying stochastic processes, 189
square-law device, 139
standard deviation, 118
stationary process, 191
stochastic process, 183
subset, 86
system, additive, 63
 deterministic, 62
 homogeneous, 63
 linear, 62, 63
 memory, 62, 63
 time-invariant, 62, 63

time autocorrelation function, 201
transformation of variables, 189, 202
transpose, 177
two-dimensional random variable, 144

Ulysses, 89
uncorrelated random variables, 147, 149
uniform distribution, 115, 124
union, 86
universal set, 88

variance, 118
variation, for aperiodic waveform, 48
 for periodic waveform, 47
vector space, 14
vectors, geometric, 13
 in mean-square estimation, 229
Venn diagram, 87

Wainstein, L. A., 239
white noise, 217, 250
wide-sense stationary, 192
Wiener, N., 4, 223
Wiener-Hopf equation, 245
wiggliness, for aperiodic waveforms, 48
 for periodic waveforms, 47
Wozencraft, J. M., 164

Zadeh, L. A., 85
Zubakov, V. D., 239